THE ESSENTIAL WRITINGS OF VANNEVAR BUSH

THE ESSENTIAL
WRITINGS OF
VANNEVAR BUSH

SELECTED, EDITED, AND INTRODUCED BY
G. PASCAL ZACHARY

Columbia University Press
New York

Columbia University Press
Publishers Since 1893
New York   Chichester, West Sussex
cup.columbia.edu

Library of Congress Cataloging-in-Publication Data
Names: Bush, Vannevar, 1890–1974 author. | Zachary, G. Pascal, editor.
Title: The essential writings of Vannevar Bush / selected, edited, and introduced by
G. Pascal Zachary.
Description: New York : Columbia University Press, [2021] | Includes index.
Identifiers: LCCN 2021033291 (print) | LCCN 2021033292 (ebook) |
ISBN 9780231116428 (hardback) | ISBN 9780231116435 (trade paperback) |
ISBN 9780231552479 (ebook)
Subjects: LCSH: Technology and state–United States. | Science and state–United
States. | Research.
Classification: LCC T21 .B872 2021 (print) | LCC T21 (ebook) | DDC 609—dc23
LC record available at https://lccn.loc.gov/2021033291
LC ebook record available at https://lccn.loc.gov/2021033292

Cover design: Chang Jae Lee
Cover image: Photo courtesy of the MIT Museum

TO JOHN MARKOFF,

JOURNALIST, HISTORIAN, TECHNOLOGICAL
SEER, AND FRIEND—A FELLOW TRAVELER TO
THE FUTURE WHO CARRIES THE PAST INTO
THE PRESENT.

"A CAPE COD YANKEE"

I'm certainly a Cape Cod Yankee. As a matter of fact, my family goes back for seven or eight generations on the Cape, and so does Mrs. Bush's. Very naturally most of the crowd was a seagoing outfit. Our grandfather took a ship to the coast of Africa as captain when he was 21. The other, I have heard, was captain of one of the three ships that first traded up the Amazon. My grandmother sailed with this latter. As near as I can make out he ran the ship and she ran the business, trading with the West Indies. She was a grand person. She lived with us until I went to college, went blind, nevertheless would not quit, and finally got killed. It might possibly be that inheritance has something to do with one's characteristics, for all of the recent ancestors were sea captains, and they have a way of running things without any doubt. So it may have been partly that, and partly my association with my grandfather, who was a whaling skipper. That left me with some inclination to run a show, once I was in it. I was not a particularly husky youngster and I had a series of difficulties: rheumatic fever, which was supposed to leave me with a bad heart but which didn't, although it left me with rheumatism which cursed me for years, so that occasionally I had to drag a leg behind me; typhoid fever, ruptured appendix, and what have you, in addition to the usual children's affairs. This had the result, for one thing, that I lost a year out of high school, and I lost a half year out of college. . . . Now of course I'm essentially a mild chap, but I'm occasionally charged with being belligerent. If there's anything in the rumor I suppose the psychologists would ascribe it to the fact that I was ill a good deal of time as a youngster. I personally would be much more inclined to attribute it (if it's true) to the fact that my great-great-grandmother came from County Kerry.

—VANNEVAR BUSH

Vannevar Bush's oral memoir, pages 1–1a, dated 1964–1965, held in the archives of the Carnegie Institution of Washington and the Massachusetts Institute of Technology.

CONTENTS

FOREWORD

I write this foreword during a pandemic. The spread of COVID-19, a novel coronavirus, throughout the United States and our networked world has created a national emergency that demands leadership by scientists, engineers, and biomedical professionals. In the United States, this emergency cuts across state and local political jurisdictions and across traditional disciplines and knowledge domains. Our nation, our world, requires technological breakthroughs, and collaboration among innovators working under intense pressure. Leadership matters. In such a time, some ask, where is a techno-scientific leader of the stature of Vannevar Bush, who during World War II mobilized the best of American and international scientists, engineers, and inventors?

Vannevar Bush (no relation to either President Bush) looms large in my life. Bush is my role model in the world of action and a source of sustenance in my scientific life. I was seven years old in 1945 when Bush's *Science, the Endless Frontier* was published, in which he presented what would become the dominant model for government funding of science and engineering research, largely through partnerships with universities. The system enabled the United States to become the global leader in science and technology in the second half of the twentieth century.

My own life in science and policy is impossible to separate from Bush's vision and legacy. I studied physics at the University of Oklahoma in the early 1960s. My academic studies and later research were funded by federal grants following a model of support for young scientists advocated by Bush. Over my entire scientific career, in research and later in university administration and service to the federal government, I benefited from Bush's vision; and I became a modest example of the iconic American Dream that so many of us in fields of science achieved in the postwar era.

In the 1990s, I took a step closer to Vannevar Bush, first when I became director of the National Science Foundation (NSF), an agency that Bush imagined and helped to create in the years after World War II. In 1998, after five years leading the NSF, I became the science adviser to President Bill Clinton in 1998, serving through the end of his term. I joined the "club" of presidential science advisers who over the decades following the end of World War II tried to fill the shoes Bush himself wore when he advised President Franklin Roosevelt on matters of military research and technology, including the design and construction of the first atomic bombs.

As NSF director and science adviser to President Clinton, I came to understand why Bush felt that scientists should speak out forcefully on important issues of public interest. While surveys show that the American people trust scientists more than most experts, they often don't understand how scientific consensus is formed and the role and methods of research and discovery. The same is true of policymakers. Echoing Bush's admonitions, as NSF director and science adviser to President Clinton, I began to stress the importance of scientists, engineers, and other technical professionals stepping outside of the Lab and engaging the public and policymakers directly on the vital importance of new knowledge and the ways in which politics and science, engineering and innovation, interact to benefit society. I came to call those who ventured into the public sphere, armed with knowledge and advice, "civic scientists." I continue to talk about "the civic scientists" whenever I get the chance and I recognize that the roots of this role model stretch back to Bush who in many ways was the original civic scientist.

The good news is that today there are many civic scientists, in America and around the world, all of whom are improving public understanding and the quality of human life, by carrying forward pieces of Bush's values and visions.

I believe that Vannevar Bush today has lessons to teach us, and insights to share, whether we are working scientists and engineers, or informed

members of society engaging social and cultural aspects of technological and scientific change. The evidence is this book, a collection of
the essential writings of Bush, carefully selected and contextualized by
Bush's biographer G. Pascal Zachary. These writings, fifty-six in total
and including Bush's signed summary to Science, the Endless Frontier,
are insightfully introduced by Zachary, a historian of technological
change and a writer on the politics of science. These diverse writings
impress me anew with the scope and scale of Bush's vision, especially in
his own field of computers, communication, and information technologies, but also across the full spectrum of science and technology. He was
a visionary but also a pragmatic operations manager, insisting that management and organization were as important for the success of complex
research projects as funding, political support, and even scientific talent. Bush knew how to talk with politicians and ordinary people alike,
including those who had little understanding of science. He helped his
case by making effective references to the concepts of "the frontier" and
"the pioneer," in order to appeal to the basic American belief in the value
of the individual and the critical importance of curiosity, freedom of
thought, and sound independent judgement.

In the mid-twentieth century, seventy-five years ago, Bush correctly
predicted that science and technology would be important to the United
States in peacetime and during the long Cold War with the Soviet Union.
He understood that the nation's ability to discover new knowledge, invent
new devices and methods, and transform them into applications, products, jobs, and wealth, creation would be challenging and require wise government policies. Given the complexity of the whole innovation process,
he has been criticized for simplifying the connections between discovery
and application in his report, but he likely felt that presenting a simple
picture was the most effective way to get his message across. Bush clearly
understood that innovation was a nonlinear process, requiring science,
engineering, and business experts to work together. He acknowledged
that innovation would be disruptive, result in tradeoffs, with inevitable
winners and losers; but that in the long term the expansion of scientific
knowledge and technological advances would benefit the American people (and so it has).

Reading these selections by Bush, many of which I'm reading for the
first time, I marvel at his breadth, depth, and prescience. He also achieved,

through his simple, direct, and clear prose, a distinctive body of writing about some of humanity's most exciting, and fearsome, new frontiers.

Neal Lane
Director of the National Science Foundation, 1993–1998
Science Adviser to President Bill Clinton, 1998–2001
Senior Fellow in Science and Technology Policy at Rice University's Baker Institute for Public Policy, Houston, Texas

INTRODUCTION

G. PASCAL ZACHARY

T o quickly grasp the significance of Vannevar Bush, recall the words of Jerome Wiesner, science adviser to President John F. Kennedy and later president of the Massachusetts Institute of Technology (MIT). After Bush's death in 1974, Wiesner wrote:

> No American has had greater influence in the growth of science and technology than Vannevar Bush, and the twentieth century may yet not produce his equal. He was an ingenious engineer and an imaginative educator, but above all he was a statesman of integrity and creative ability. He organized and led history's greatest research program during World War II and, with a profound understanding of implications for the future, charted the course of national policy during the years that followed.[1]

To many now, a half-century since his death, Vannevar Bush is best known for his formative role in the rise of computing and for conceiving of, and helping birth, the National Science Foundation (NSF), the leading supporter in the United States of basic scientific and technological research. In his prime, Bush (1890–1974) was better known as the organizer of the Manhattan Project and as the science and military technology adviser,

---

1. Jerome Wiesner, "Vannevar Bush," *Biographical Memoirs*, vol. 50 (Washington, D.C.: National Academy of Sciences, 1979), 89.

first to President Franklin D. Roosevelt during World War II and then to Roosevelt's successor, President Harry Truman, immediately following the war. Joining with other presidential advisers, Bush recommended that Truman drop atomic bombs on Japan in August 1945. Bush then became, for a time, a leading public voice on the perils and necessities of nuclear weapons. At the onset of the Cold War with the Soviet Union, and continuing throughout the 1950s, he was one of the most vocal and influential Americans on the subject of science, military technology, and national security. Even in the 1960s, Bush gained attention for his criticisms of the Apollo space program and, in particular, of the pursuit of landing men on the moon, which he viewed as "spectacular" entertainment rather than meaningful science.

While comfortable wielding his influence in private conversations in the corridors of the Pentagon, the White House, and elite universities and foundations, Bush also relied on the power of his pen to persuade others. He wrote well in plain English, and he wrote often: essays, speeches, public testimony, lengthy memos, and short private letters. He published significant pieces in leading newspapers and magazines, and he shared his views through memos to powerful figures, privately circulated essays to the classified government community, and in a popular and widely read book titled *Modern Arms and Free Men*, concerning the effect of nuclear weapons and emerging technologies on military strategy, national security, and the process of innovation in a democracy.

Yet virtually all of Bush's writings are out of print, from *Modern Arms* to his rambling memoir, *Pieces of the Action*, and even a thin book of essays with the arresting title *Science Is Not Enough*. Moreover, there exists no anthology or edited volume of his writing across the full range of his expertise and concerns. *The Essential Writings of Vannevar Bush* collects together in a single volume—for the first time—some fifty-six selections of the most relevant and historically important of Bush's writings. As the sole biographer of Vannevar Bush, I have drawn on my extensive knowledge of his life and times to select and curate these writings from the many candidates in his literary corpus. Many of these selections have never been published before and can only be found in Bush's sprawling personal papers at the Library of Congress, scattered in the National Archives, or in small collections maintained by MIT and the Carnegie Institution of Washington, the two organizations where Bush spent nearly his entire working life.

Bush valued clear and effective writing, and he wrote often as a way of exploring his personal interests, especially on the management of information and of increasingly important large organizations. Many of his most significant pieces advocate for specific policies and initiatives in science, engineering, and public policy. To Bush, writing was an essential activity, a professional obligation if not a personal passion. To write, in Bush's mind, came from a sense of civic duty. In Bush's first textbook, entitled *The Principles of Electrical Engineering*, which he coauthored and published in 1922 with a senior colleague William H. Timbie, a senior faculty colleague at MIT—he emphasized to aspiring engineers the importance of effective and persuasive writing. As Bush and Timbie explained in their opening chapter on "fundamental considerations," the success of any engineer's plan often depends on good writing. "Once the plan has been decided upon, he must convince his superiors that the plan should be carried out," they wrote. "This convincing requires that the engineer write clear, brief English, which adequately and concisely conveys the meaning in a convincing way. Good proposals have been turned down because the engineers who drafted them could not present them in convincing form."

Bush gave this advice based on his own experience. Consider Bush's first patent application, in February 1912. Patent applications are rarely considered a form of literature, but in Bush's, we can see the germs of his later style: precise, clear, concrete, and confident in tackling technical descriptions in plain English. He also tries to identify, and sometimes succeeds in identifying, the stakes of his technical achievement.

To all whom it may concern,

Be it known that I, Vannevar Bush, a citizen of the United States and a resident of Chelsea, in the county of Suffolk and the State of Massachusetts, [has] invented certain new and useful improvements in profile tracers, of which the following is a specification.

This invention relates to an instrument adapted to be used in surveying for drawing a line upon a record sheet representing the elevations of successive points in a line extending across a strip of the ground. In surveying work, particularly surveys of railroads and continuous lines, it is customary to prepare profiles of the country along lines to the transverse direction of the road at suitable intervals.

The purpose of the invention is to provide an instrument which will trace the superficial outlines of such profiles quickly and with approximate accuracy, which is automatic and capable of recording elevations of all points along the line on which the profile is taken with greater exactness than is obtainable by the usually practiced methods of surveying, except by expenditure of much time and labor.

My main objective in developing the present invention has been to secure accuracy.

A secondary objective has been to produce an apparatus which is not only accurate but is exceedingly simple in construction and has slight liability to derangement.

The manner in which I have accomplished these objects is disclosed in the following specification, wherein I have described and illustrated an instrument embodying the essential principles of the invention.[2]

The Patent Office approved his application on December 31, 1912. He was only twenty-two years old.

Bush's early talent for inventing opened a path for him in academic science, which in turn led to his own organized research and then to the management of research by others. Throughout his life, his interests and activities varied extraordinarily. In the 1920s, he cofounded the electronics company Raytheon. In the 1930s, he designed what were then considered to be the world's most powerful electro-mechanical computers. And during the 1940s, from his office near DuPont Circle in Washington, D.C., he directed military advances in radar, precision bombing, and atomic weapons—all crucial to the outcome of World War II and to shaping American responses to the postwar challenges. The writings I selected for this volume cover a span of forty years, from 1929 to 1970, and explore such varied topics as war and technology, engineering, discovery and innovation, knowledge and national security, the origins and future of nuclear weapons, the politics of science, the role of technology

---

2. William H. Timbie and Vannevar Bush, *Principles of Electrical Engineering* (New York: John Wiley & Sons, 1922), 8.

in public life, and the development of computers and digital networks and their role effect on human memory, consciousness, and purposeful thought.

Bush wrote simply, directly, and usually concisely and clearly. There is a stylistic unity to his writings that somehow survives across the expansive subjects that occupy and at times possess him. I found it impossible, however, to organize Bush's writings by theme or style or subject. The selections are presented chronologically, from the date of first circulation (in the case of speeches, memos, and letters) or the date of publication. The advantage of chronological order is that readers can identify, often quickly, how Bush's ideas and concerns cross-pollinated one another and were shaped by and influenced the times and conditions during which he wrote.

The selections begin with the preface Bush wrote for his first book, a technical treatise, or textbook, published in 1929. Bush was thirty-nine years old, entering the prime of his career as an engineering educator. The last selection is his final reflection on computers, consciousness, and the human condition, published in 1970, some five years before the advent of the personal computer. Bush visits this subject a half dozen times in his works, most notably in an excerpt from his often-cited essay, "As We May Think" (chapter 21) and a vital antecedent, "To the Things of the Mind," an unpublished essay he wrote in 1941 (chapter 15). In "The Peak Wave of Progress in Digital Machinery" (chapter 36), he amply justifies his standing among contemporary computer and software designers as an early prophet on the information age whose main insights on how information is stored, organized, retrieved, and amplified remain influential to this day.

Bush is best known for his introduction to a report titled *Science, the Endless Frontier*, which he delivered to Truman in July 1945 at the end of World War II, and the summary is included in its entirety as chapter 23. To this day, *Science, the Endless Frontier* shapes thinking about science and society and the government funding of research. But much of Bush's finest writing remains unknown, unavailable, and unappreciated, yet worth our attention.

Henry James, the great novelist and literary critic, shrewdly identified the reasons to revive a neglected writer. "In every attempted resuscitation of an old author, one or two things is either expressly or tacitly claimed for him," James wrote in an essay published in 1866. "He is conceived to possess either an historical or an intrinsic interest. He is introduced to use either as a phenomenon, an object worthy of study in connection with a particular phase of civilization, or as a teacher, an object worthy of study in himself, independently of time or place. In one case, in a word, he is offered to us as a means; in the other case, he is offered to us as an end."[3]

For historical significance and as "an object worthy of study" himself, Vannevar Bush demands reading today. His defense of J. Robert Oppenheimer, unfairly accused of disloyalty despite his leading role in the Manhattan Project, and Bush's own secret effort to halt the first test of a hydrogen bomb (see chapters 37 and 38) are only two lesser-known examples of writings that illuminate and enrich our understanding of the American past in a global context. Bush was a leader of scientific organizations and an engineer and mathematician of significance, and his writings show how the imagination of a unique American creator infuses meaning and value into the permanent revolutions wrought by techno-science. He presents such techno-scientific complexity at a human scale, and he writes with a clarity, precision, and lyricism typical of poets but rarely found in scientists and engineers. With infrequent pomposity and scant self-importance, Bush brings literary values and persistent philosophical concerns to the task of making sense of the sometimes dizzying transformations in the human-built world. And he does so, invariably, by gazing in one instant forward and backward in the next.

Bush lived in a print era. He rarely appeared on radio or television (though once, in 1956, he starred on Edward R. Murrow's popular *Person to Person* broadcast). Print was his medium of choice, though he wasn't precious about writing. He wrote purposefully and to make a point. He wrote longhand, and he typed, though not often. He dictated his words to others, and he talked into a tape recorder and later revised transcriptions. While writing is not the chief measure of Vannevar Bush, his written words today best capture his legacy.

Today, many writers spend their time making sense of emerging technology and science. Entire literary careers are built on exploring the subjects of bio-medicine, computing, the environment, and the human-built world. This was not so in Bush's prime years, from the 1930s to the end of the 1960s. While nuclear weapons spawned a great deal of writing during

---

3. Henry James, review of *The Works of Epictetus*, by Thomas Wentworth Higginson, *North American* Review 102, no. 211 (1866): 599.

and after World War II (including the essay that Bush coauthored with Harvard president James Conant in 1944, included as chapter 18, on how quickly other nations might shatter the American "nuclear monopoly in the making"), literate writing about technological change was rare until the 1970s and only flowered after Bush's death in 1974. In his writing prime, Bush was a rarity in his zeal for translating complex techno-scientific ideas and trajectories into plain English. For the quarter century from 1945 to 1970, he ranked among a very few scientists and engineers with a flair for proving clear explanations and judgement about tech trends and trajectories. His writings in the 1960s on the Apollo program, especially his letters to NASA director James Webb (chapter 48), are models for clear and compelling explanations of complex socio-technical situations. Bush was one who could do this and do it with a certain felicity and style. His privileged place and his role as agent and sponsor of revolutionary technologies was partly why. He was among the first people in history to think elegantly and intelligently about the effects on humans of nuclear weapons (mostly negative) and on computers (mostly positive).

In the mid-twentieth century, Bush ranked among a handful visionary technologists who thought deeply about the effect of innovation on American life and America's new role in world affairs. At a time when few scientists and even fewer engineers wrote for general audiences, Bush wrote a best-selling book on technological change, war, and national security; he published essays in leading popular magazines, most notably the *Atlantic Monthly*, whose legendary editor Edward Weeks once described an essay of Bush's that he edited to be as significant as Emerson's "The American Scholar." Like Emerson, Weeks insisted, Bush's writing advanced a "new relationship between thinking man and the sum of our knowledge."

Intelligence and foresight are only two of the virtues to be found in Bush's literary output. There's also the pure enjoyment of reading him, and the element of surprise in his exposition. Curiosity drove him. His father Perry, a minister, wrote poetry and self-published a volume of verse, which his son kept close. Bush enjoyed reading his father's sermons and sometimes in his own writing he veered into sermonizing (as in chapter 10, "The Qualities of a Profession," and chapter 53, "The Art of Management"), where he advanced the high-minded, humanistic idea that engineering and organizational leadership were forms of "ministry to the people," that should be "exercised with pride and dignity."

But Bush was not perfect, and as a technological seer, he made errors. He downplayed for too long the prospects for guided missiles, and he derided the possibility of "push-button war," which would send nuclear-armed missiles at the United States with the touch of a Soviet button, to his later embarrassment. While criticizing the Apollo space program for its naked political theater, he failed to grasp that the Space Race created popular support for science and engineering. He remained in the thrall of the immense achievements of World War II, when a global emergency suspended normal government processes and allowed a talented elite, usually protected from political scrutiny, to make sweeping decisions with little to no oversight. The notion of politics today as a big tent that demands pluralism and inclusion would seem unattractive, even alien to Bush. His preference for private decisions and shielding official policy differences on science and technology to a select few technocrats is simply unacceptable today, not to mention unwise. Along with his elite peers at the top of the science and engineering enterprise in the United States from the 1930s to 1950s, Bush took virtually no steps to open jobs, research opportunities, or leadership positions for talented women and African Americans.

Bush's vision for computing was limited by his weak grasp of the potential of digital electronics to revolutionize computer processing speed and data storage. After World War II, he never experimented with integrated circuits, invented in 1947 at AT&T's Bell Labs. Bush died only a year before the first personal computers were offered for sale to the public, but in the years prior to the advent of the PC, in the mid-1970s, he had little or no contact with the new generation of young computing enthusiasts. However, even as he ignored digital electronics, he promoted a vision of automating thought and analysis through computation and advocated the mass acceptance of computers by individual scholars and scientists, professionals, and creative workers. In contrast to the popular view of computers as forces for routinization and dehumanization, Bush countered with a human-centered vision of "augmentation," or enhancement of human thought and creativity, through the embrace of desktop computers as a "tools for thought," in the words of historian Howard Rheingold. "There will always be plenty of things to compute in the detailed affairs of millions of people doing complicated things," Bush wrote in his landmark "As We May Think" essay, in which he envisioned "wholly new forms of encyclopedias . . . ready made with a mesh of associative trails running through them," searchable on desktop devices with familiar keyboards. Bush even imagined a future occupation that today people think of as "data mining," presciently predicting the rise of "a new profession of trail blazers, those who find delight in the task of establishing useful trails through the enormous mass of the common record."

———— ⌬ ————

By any definition, Bush was an American original. A visionary electrical engineer in an era of mass electrification, he was also an applied mathematician who attacked nearly every socio-technical problem known to humanity, from controlling weapons of mass destruction and solar energy to the automation of thought, the mysteries of bird life, and the vagaries of car engines. He came to prominence while a professor at MIT for his designs and conceptions of what in the 1930s were the most powerful electro-mechanical computers in the world. He was more, though, than a forerunner of the computing revolution. The only son of a New England minister, he was a polymath by inclination and worldly wise. In 1939, he left MIT to lead the Carnegie Institution of Washington, America's leading private philanthropy in support of scientific research. The job of president was a boon, paying the princely sum of $25,000 a year. The Carnegie Instituiton of Washington maintained laboratories in California, New York, and Washington, DC, supporting scientists to do research full-time and freeing them from teaching and administrative duties.

After Andrew Carnegie, the philanthropist who founded the institution, died in 1919, leadership shifted to the institution's influential board, which included former president Herbert Hoover, the celebrated aviator Charles Lindbergh, and Frederic Delano, uncle to President Roosevelt. The chairman of the board, W. Cameron Forbes, hailed from a wealthy Boston family and knew Bush at MIT, where, along with his distinguished research, he had served as second-in-command to the university's visionary president, Karl Compton. Delano, who was Roosevelt's private liaison to the nation's science community, personally wrote to Bush on behalf of the board with the formal job offer.

Bush officially took the reins at Carnegie on January 1, 1939, and he immediately set to work bringing broader perspective on scientific research to the nation. More importantly, as president, he gained invitations to national policy committees on science and engineering. He quickly gained a position on the National Advisory Committee for Aeronautics, a federal agency responsible for innovation in aviation. The position introduced Bush to competition with Germany, whose aircrafts were far superior to those designed and built in the United States. He quickly advocated for the government to expend "much more effort to our own aviation development if we are to keep pace."

By the fall of 1939, Germany conquered neighboring Poland, and by May of the next year, they had invaded France and threatened to do the same to Britain. Alarmed by German aggression, Bush spoke to a gathering on military aviation enthusiasts on May 29. This was the first time he displayed his enduring, single-minded vision for an American nation sustained by technological innovation. "For war or for peace," he said, "we must leave no stones unturned in research."

Bush matched forceful rhetoric with a program for action. He conceived of a way to rapidly mobilize American scientists and engineers on behalf of national security and asked Frederic Delano to connect him to Roosevelt. Delano delivered. Bush first met Harry Hopkins, perhaps Roosevelt's closest aide, and brought him a short memo describing a plan for a coordinating committee, responsible to the president, that would contract with universities and industrial labs to perform research useful to the army and the navy. On June 12, Bush—joined by Hopkins—met Roosevelt for the first time. Roosevelt approved the formation of the National Defense Research Committee, which Bush would lead.

Bush and Roosevelt worked closely together, and Bush became his unofficial science adviser. Roosevelt chose Bush to organize research of the first atomic bombs and relied on Bush to brief him on the Manhattan Project. In turn, Lieutenant General Leslie Groves and physicist Robert Oppenheimer—together the twin pillars of the A-bomb project—reported directly to Bush and his alter ego in atomic research, Harvard president James Conant. (Bush and Conant's seminal memo, "Salient Points Concerning the Future of Atomic Bombs," forms chapter 18; the memo accurately predicted future atomic competition among nations for atomic supremacy.)

In November 1944, when American victory seemed inevitable, Roosevelt, at Bush's insistence, formally asked him to create a plan for government support for science and technology after the end of the war. After Roosevelt's death in 1945, Bush delivered the report to Harry Truman, his successor. For a short time, Bush served as Truman's science adviser, but by 1947, with Cold War tensions rising with the Soviet Union, Bush lost his access to the White House, though he remained an important private and public voice in policy discussions about nuclear weapons, the course of military technology, and the federal support for civilian science. On this last matter, Bush helped conceive of a national foundation for research. The NSF came into being in 1950.

One can gain a sense of Bush's surprising breadth and staggeringly diverse interests and activities by looking at a single month of his life: July 1945.

On July 5, Bush sent a report to Truman on future government support for science in which Bush called for, among other things, the creation of a national research foundation, which by 1950 would morph into the NSF. Months in the making, the report, shaped by Bush and called "Science, the Endless Frontier," asked Americans, weary of war and fearful of a return to economic depression, to explore a new frontier, a frontier of knowledge, scientific discovery, and socially useful innovations. "The pioneer spirit is still vigorous within this nation," Bush wrote to Truman. "Science offers a largely unexplored hinterland for the pioneer who has the tools for his task."

By calling for a new federal agency to support science, Bush sought to link discovery to the return of American peace and prosperity. He also sought to ensure that the pool of scientists in the United States, greatly enlarged by European refugees from fascism and war, would be replenished by home-grown talent. Bush further envisioned that federal support for science and engineering would be chiefly carried out by universities and through the technique of the "research contract," which Bush essentially invented during the war. The contract paid scientists for effort, not necessarily out-comes delivered, recognizing that research is inherently unpredictable.

Eleven days later, on July 16, 1945, Bush observed the Trinity test in New Mexico. It was the first atomic explosion on earth. Early in World War II, at the request of Roosevelt, Bush had organized the Manhat-tan Project, serving as the official White House liaison to Groves and Oppenheimer. After Roosevelt's death, Bush advised Secretary of War Stimson and the new President Truman on use of the new weapon against Japan.

Later that same month of July 1945, Bush published a long essay in the *Atlantic Monthly* that garnered great attention and even a cover story in *Life*. To this day, Bush's article, "As We May Think" (chapter 21), is viewed by the leaders of the digital industry—or Big Tech, as it is sometimes called—as anticipating and predicting the personal computer revolution of the 1970s, the search engine and Web of the 1990s, and the social media transformation of the early twenty-first century.

July 1945 shows how even while managing the largest research and development effort during a global war, Bush continued to think ahead. Not only about weapons of mass destruction and their effect on civilization, but also about how computers would alter human life for decades to come. He was especially intrigued by the possibility of computing machines easing what Bush called "mental drudgery." With great prescience, Bush described how the desktop computer would become a memex, or memory extender, and how information storage would allow us to organize our ideas and records according to "association," which would later be called hyperlinking, the technique that links data points and makes fast Web searches possible.

Bush hoped to solve the emerging problem of "information overload," but he also saw—almost alone in his generation—that computers could manage all of the world's knowledge and thus "lighten the burden on man's mind, as earlier [technological] developments lifted the load from his muscles." This could expand human creativity and knowledge on a vast scale and bring about a kind of digital immortality to humanity. "The inheritance from the master will become, not only his additions to the world's record, but for his disciples the entire scaffolding by which they were erected," Bush wrote. "Each generation will receive from its predecessor, not a conglomerate mass of discrete facts and theories, but an interconnected web which covers all that the [human] race has thus far attained." Not masses of information, but coherent bits of knowledge and patterns of understanding are what Bush foresaw computer-aids would bring.

Bush's vision inspired generations of computer designers. Douglas Engelbart read "As We May Think" in the 1940s and went on to invent the first computer mouse and first email system in the 1960s. Larry Page and Sergey Brin, cofounders of Google, were introduced to Bush's ideas as graduate students at Stanford University in the 1990s. In an eerie forerunner of Google's search engine, Bush imagined "an interconnected web" that would accelerate the advance of civilization by freeing the human mind to concentrate on solving the hardest problems quickly. This happy outcome, Bush imagined, depended on "the task of selection," which today is known as "search," the basic tool for managing the "web," a term Bush may well have coined (but didn't profit from).

To be sure, Bush didn't foresee the hegemony of digital computing, because he held too fast to the electro-mechanical technologies of his era. Yet Bush fully grasped the deep consequences of the radical miniaturization of information storage and how new innovations would alter the status of the printed book. "The *Encyclopedia Britannica* could be reduced to the volume of a matchbox," he wrote. "A library of a million volumes could be compressed into one end of a desk." Moreover, Bush envisioned that miniaturization would spawn appealing innovations in information media. Anticipating Wikipedia's creative destruction of large, printed encyclopedias, Bush predicted, "Wholly new forms of encyclopedias will appear, ready-made with a mesh of associative trails running through them, ready to be dropped into the memex and there amplified."

While the month of July 1945 drew on the full range of Bush's talents, the atomic explosion on Hiroshima on August 6 stoked his darker sensibilities. American science and engineering, with all of its promise, also produced a sense of instability and anxiety. Suddenly religious and social thinkers, as well as ordinary workers and managers, soldiers and statesmen, openly worried whether science and human survival were compatible or tragically at odds.

Bush sought to counter pessimism with a gritty pragmatism and an implicit belief in progress. He insisted doing science was essential to human thriving, and his book, *Modern Arms and Free Men* (excerpts presented in chapters 28 and 29), outlined how. During these early years of the Cold War, he advocated (unsuccessfully) for placing the direction of most if not all federally funded research—civilian and military—under his proposed research foundation. Instead, Congress and the president created the Office of Naval Research and the Atomic Energy Commission to direct and administer military innovators. Bush's insistence that his research foundation have no direct political control also proved unworkable, which delayed Truman's approval of a modified, civilian-only NSF until 1950.

Setbacks in public policy jolted Bush enough that in 1961, he admitted that the role of scientists in the political process was like a baseball team playing a road game in "the other fellow's ballpark." In a bow to the talents of the political class, Bush confessed, "Scientists today are privileged to participate [in politics] to an extent never before true in this country, and it is certainly incumbent on [scientists] to understand and indeed to sympathize with the local ground rules which govern the ball park in which they are now exercising important influence."

Bush cared deeply about civil-military relations and thought that this central tension in American public life was fundamentally misunderstood. His most important statement on helping the military maintain its edge in new weapons and tactics is found in "A Few Quick," an essay privately circulated by Bush in 1951 that was highly influential among American military leaders. This essay, published here for the first time as chapter 32, establishes Bush as one of the few technologists of mid-century America who thought deeply about the importance of speed in delivering new innovations, the necessity of freeing innovators from bureaucratic constraints, and the challenges of matching emerging military technologies to shifting requirements in the real world. While American intellectuals at mid-century tended to reflect on the challenges of civilian control of the military, Bush obsessed over how talented civilians, especially those with expertise in science and engineering, management, and organization, might directly influence the world of the soldier—and to do so while embedded within military structures and even during the conduct of war. Looking back on the Allied victory in World War II, Bush found large and enduring value in close collaboration between civilians and the military, warriors and inventive technologists.

If the political theorist Samuel Huntington worried about civil-military relations from the standpoint of democracy, Bush viewed the problem of national security as the task of organizing the military around new weapons, new tactics, and emerging innovations. Bush wanted the generals and the admirals to see around the bend in the river, to anticipate where combat might go, and where threats might emerge. For Bush, "soldiers and scientists in partnership" was not an empty slogan, but the principle from which sound military planning and operations would arise.

Bush's writings inspired civilian innovators too. He was fascinated with entrepreneurship long before the subject became fashionable. His 1939 testimony to Congress, the text of which comprises chapter 9, contains many insights that sound strikingly contemporary and prefigure the rise of venture capital, emerging information technologies, and the risk-taking entrepreneur. During the Depression of the 1930s, Bush anticipated, while not explicitly espousing, the concept of creative destruction, which was conceived in the same milieu by Austrian economist Joseph Schumpeter. Like Schumpeter, Bush worried about the losers in the innovation process and how big companies might stifle innovations in favor of protecting and profiting from existing technologies. Bush celebrated the individual inventor and the business-minded technologist whom Schumpeter labeled "the entrepreneur."

Bush knew from the Manhattan Project that what seemed to flow inexorably from a scientific breakthrough—say, the achievement of an understanding of the physics of atomic fission—also depended on an industrial organization of enormous scale. Experience in war, in short, taught Bush to straddle the worlds of science and engineering, the small entrepreneur and the giant company, and the academic world of "pure" discovery and the pressurized pursuit of making practical inventions work in the field.

In his teaching, Bush sought to imbue junior faculty as well as doctoral students with a passion for making a difference in the world through capitalist exploitation of discovery and invention. At MIT, where Bush worked in the 1930s, he embraced an interplay between industry and academia. He pushed professors to turn their knowledge into action and even into commercial gain when possible. Frederick Terman, whose PhD Bush supervised at MIT, moved to Stanford University in Northern California where he promoted academic collaborations so vigorously that he became known as the father of Silicon Valley. Another doctoral student Bush advised, Claude Shannon, researched practical problems of telephony for Bell Labs. Shannon became an epic force in digital communications, applying Boolean logic to new digital switches and vividly demonstrating that engineers and mathematicians can, as much as physical scientists, create new knowledge. At the peak of Shannon's impact, Bush joined AT&T's board of directors. He also served for a time as chairman of Merck, the pharmaceutical leader.

We can see now that Bush did more than pursue research and discovery. While the importance of use-inspired research stands among his vital legacies, Bush remained preoccupied with human curiosity, and he was struck by the parallels between science and art, engineering and music, and the human search for knowledge and understanding for its own sake. On the tenth anniversary of Hiroshima in 1955, Bush provided readers of the *Atlantic* with his answer to the eponymous question: "Why Do We Pursue Science at All?" (chapter 42). His closing words resonate with us still:

> The simple creed of service . . . suffices for the day's work of many a scientist. Yet there is joined with it a deep conviction, a faith if you will, which for many a [scientist] furnishes motivation and satisfaction . . . This is the conviction that it is good for [hu]man[s] to know, that striving for understanding is his [and her] mission. We [scientists and engineers] are embarked upon a great adventure, and it is our privilege to further it. Even though at times the box that is opened be Pandora's, even though there are both good and evil in what we learn, it is our duty and our calling to extend [humanity's] grasp of the universe in which [we] live. . . . By this process, of beginning to understand, we have made such progress as we have. Though the path be thorny, this is still the way in which we should proceed if we would finally emerge from darkness into the light.

I will stop here. As Bush's only biographer, I have long felt that collecting his writings would be an obligatory act of scholarship—an act of devotion, perhaps, to Bush himself—and my own search for understanding. But now, after editing this formidable book of his writings, I must also confess I found joy in it. And now I have said enough. There is no substitute for reading Bush in his own words.

Each of the fifty-six selections carry a power and authority that cannot be summarized, condensed, or neatly explained. To orient the reader, I have provided a brief introduction to each selection—for context and sometimes to explain seemingly puzzling or confusing aspects of Bush's attitudes or preoccupations. I have only one apology to make on his behalf. I regret his frequent use of the words "man," "mankind," and "men," when he ought to have written "human," "humanity," "humankind," and, most obviously, "men and women." We need not forgive Bush his flagrant gender bias in order to benefit from his wit and wisdom on the intersection of science and society, engineering and human prospects. In our time, when writing about science and technology is itself a massive industry, and when scores of writers undertake it, Bush brings us back to a time when few Americans thought deeply about science and discovery, engineering and enterprise. Present at the creation for so much, Bush shaped how we have thought and written about the technological foundations of the everyday human-built world. We live in this world today, and it continues to elude our complete control and understanding in ways that would not surprise Bush at all.

EDITOR'S NOTE

The selection of Bush's writings, and their chronological presentation in this volume, are wholly my responsibility. I applied a few broad criteria in my choices, and I want to share the criteria with readers, albeit briefly.

First, I weighed the historical importance of every selection, especially with regard to insights into the Cold War rivalry with the Soviet Union, the evolution of nuclear weapons, the state sponsorship of innovation in the United States post-1940, the emergence of visions of and aspirations for a revolution in computing and information management, the role of engineers in society as well as their changing self-conception as creators of new technologies, new knowledge, and new ways of life.

Second, I assessed the literary style and value of a selection, especially with regard to Bush's own project of creating a literature of engineering and a manner of writing about technological change where he sought to balance novelty and significance, clarity and technical depth.

Third, I considered the relevance to contemporary debates around the politics of science, the business of technology, and the economics of research, discovery, and innovation.

If a potential selection "scored," in my eyes, highly on a single metric, I included it. Usually, a selection merits consideration on all three parameters. For each selection chosen, I provide a brief introduction. My aim is to highlight significant points in the selection and to put the selection into some historical, technological, cultural, or intellectual context.

I occasionally trimmed selections for readability or to remove obscure references that were not relevant to Bush's main points and would require too much explanation to retain without distracting or confusing the reader. These deletions are indicated by ellipses (. . .). I changed punctuation, in some cases, to match contemporary practice and enhance readability. At times, I substituted in brackets the words "human" or "humanity" for Bush's preferred terms of "man" and "mankind" in order to conform to contemporary usage and convey more fully Bush's meaning. In Bush's writing, his references are invariably to men; he refers uniformly to "his" or "him" and never to "her." While unfortunate and outmoded, his male-centric mentality reflects the times in which he lived, his unacknowledged gender prejudices, and his social blind spots.

The source of a selection is always found in the note on the first page of each selection. I have sometimes chosen an earlier draft of an essay published by Bush because the antecedent provided more historical context and sometimes presented an argument or a perspective more effectively.

THE ESSENTIAL WRITINGS OF
VANNEVAR BUSH

1

PREFACE TO *OPERATIONAL CIRCUIT ANALYSIS* (1929)

In Bush's first textbook as sole author, he writes in the manner advocated by the late Jacques Barzun, whose classic guide to writers is entitled Simple & Direct. *Bush followed this literary dictum from the very start of his publishing career.* Operational Circuit Analysis *achieved some notice (and went through at least five printings and two editions, the last in April, 1948) perhaps because of the prominence of his own scholarly research and academic position at MIT. A reviewer in the* Bulletin of the American Mathematical Society *declared Bush's prose to be "pleasing" and, while "an elementary text," the book well represented its technical subject from "the standpoint of the engineer." The reviewer added, "one might occasionally wish for somewhat more precision" in his expression. Throughout the text, Bush's writing is animated by the values of simplicity and directness, which also shaped his approach to engineering problem-solving. He also shows enthusiasm for serving in the role of broker or translator of concepts and procedures from one knowledge domain to another. What's most striking in the preface is Bush's insistence "that I write as an engineer." Here Bush furtively gives birth to a literature of engineering, by and for engineers, and he goes on to advance a singular ambition, probably never before expressed in American letters, that he plans "to do [his] part to bridge the chasm that so often separates the engineer and the pure scientist." And Bush would spend the rest of his writing life building that bridge.*

PREFACE TO *OPERATIONAL CIRCUIT ANALYSIS*

There is now considerable literature on the subject of operational analysis of circuits, which has developed and extended the methods introduced by Oliver Heaviside [the English engineer and mathematician, who had died four years earlier] in his classic electromagnetic theory and electrical papers. Yet there is no text which gathers together the various parts of the subject in a form available to engineers and physicists. This the present treatment aims to accomplish, without undue emphasis on any one aspect or point of approach.

The operational method is very powerful, and it can be just as rigorous as the classic analysis on which it is based. It should be, and undoubtedly will be, increasingly used by all who deal with circuit transients. This contemplates not only the circuits of electricity, but also those of acoustics, mechanics, thermics, hydraulics, and so on. The object of this text is an exposition and substantiation of operational methods in a form applicable to all sorts of circuit problems. Illustrations are given freely and these are often electrical; but this is not a collection of interesting electrical problems, nor a treatment of transmission lines and cables.

It is entirely possible for a computer to perform the algebraic work necessary for the symbolic solution of alternating-current networks in the steady state without any grasp of the philosophy of the symbolic treatment or of the mathematics of differential equations on which it is based. The operational method bears much the same relation to transient problems that the symbolic method bears to steady-state problems. Both are shorthand processes based on classic circuit analysis. It is entirely possible to utilize the operational method on specific problems without the slightest idea of why and when it does or does not work. This, however, is computation and not analysis. In order to extend the operational method, to apply it freely and safely, to save time by the employment of its directness and simplicity on new problems, it is necessary to have at least some grasp of the classic mathematics for which it is a working tool. . . . [But] a little mathematical knowledge can be a dangerous thing, and the use of operational methods except for pure computation should be accompanied by an appreciation of the logic of more than simple algebra.

Preface, dated January 1929, from the original edition of Bush's textbook on circuit analysis, published by John Wiley & Sons.

Finally, it should be emphasized that I write as an engineer, and that I do not pretend to be a mathematician. I lean for support, and expect always so to lean, upon the mathematician, just as I must lean upon the chemist, the physician, or the lawyer. It is certainly incumbent upon me, however, to learn the language and as much as I can of the processes of thought of the mathematician, to do my part to bridge the chasm that so often separates the engineer and the pure scientist. I write in the hope that this will be the attitude also of these of my readers to whom this text will be useful. Perhaps it will not be uninteresting to the mathematician to note an engineer's approach to a problem with this situation in mind. . . .

2

THE KEY TO ACCOMPLISHMENT (1932)

By the age of forty-two, Vannevar Bush had undeniably "made it" in America. He was well established as a pioneer in the infant field of computing and a leader of the powerful electrical-engineering faculty at the Massachusetts Institute of Technology, the foremost engineering university in the United States. He had also launched a successful electronics company, Raytheon, with a classmate from his undergraduate years at Tufts University. In this inspirational essay for a publication of the U.S. Institute for Textile Research, Bush displays an uncanny sense of how discovery and the search for new knowledge often work best when harnessed to the pursuit of utility, or practical aims. In this early example of his views on innovation, Bush also shows glimpses of his rare talent for combining technical subjects with social values and applying scientific perspectives to business and commercial issues. He also sounds a theme which will come to characterize his public persona and his ample achievements in the name of the American people. "The key to accomplishment is research," he writes. "It need not be complex, in order to be useful, but it certainly should be intelligent." He adds, presciently, "In our modern tempo, that industry is in danger which is in a static state" and research "is a necessity" if only as "a defense against encroachment from without."

THE KEY TO ACCOMPLISHMENT

Considering the fact that he has been at it for only a few thousand years, man makes rather good textiles. He even is able to make acceptable textile fibers. Compared to nature, however, he is still an amateur with much to learn. Of course, nature has carried on its development over some hundreds of thousands of years, and man cannot wait that long. Fortunately he doesn't have to. He can accomplish in a decade results which would require hundreds of years by empirical methods or thousands of years by processes of evolution and natural selection. This he can do in one way only: by research, intelligent and intensive research, vigorously supported and patiently followed up in production. The purpose of this [essay] is to review some of the most important contemporary research in the textile field [as a means of understanding the value of research in general].

The first textile research worker sat on a log by a stream and plaited rushes to form a mat. His name is lost to us, if he had one, but his ingenuity was in first-class working order. It probably required many generations before the mat became a basket, and the basket acquired a handle; but the basic idea of interwoven textiles was there born. We still seldom depart from the fundamental scheme of interlacing fibers to form a flexible sheet of material. For some reason or other nature does not seem to have followed out this same line of development; but still we must admit that nature has made better clothes than we.

I have always envied the duck. He can dive under water and come up dry. Yet his coat is pervious to air as it should be for his good health, and it fits beautifully. The duck looks comfortable in his waterproof garments on a hot day, but the only raincoat I ever bought was hot, or it wasn't waterproof, and it leaked at the neck when the rain drove horizontally. The Mongolians made a pervious felt that shed rain well, and Caesar thought well enough of it to adopt it. The duck can turn his head in any direction, and yet his covering on his neck lies marvelously smooth and sleek. Columbus found the Indians of Central America using feather garments; but even with their example, we still do not emulate the duck, nor do we look natty

Vannevar Bush originally wrote this essay to introduce a book, published in 1932, by the U.S. Institute for Textile Research, *Textile Research: A Survey of Progress*.

or comfortable in the rain. It is certainly true that Solomon in all his glory was not arrayed like one of our modern women—not a queen but a woman of the people. We have progressed. Yet the lily can still exhibit more pleasing finish and coloring, to my way of thinking; and the grass of the fields presents me with more alluring gradations of greens and browns than I find in neckties. Moreover, the grass takes on more and more attractive hues after long exposure to the sun and rain, and neckties do not.

This is not intended to be a criticism of the textile research worker who, after all, has been studying his subject for only a short time and who has produced some marvelous fibers and fabrics. Rather we should note that there is no field of human endeavor in which so much ingenuity and resourcefulness has been shown as in textiles, or which has brought more of lasting benefit to mankind.

When civilization meets a crisis, a world war and a depression, we are reminded forcefully that there are three fundamental needs of man: food, shelter, and clothing. Transportation, communication, and the rest are contributory rather than basic. There are three primary things we cannot do without, at least in this climate. The supplying of food has given rise to a large number of diversified industries. The furnishing of shelter, again, is in the hands of many groups of widely different types. But the production of clothing, while it involves many branches and is in the hands of a large number of small manufacturing units, is in its interests much nearer to being a close-knit homogenous whole. With its importance, the uniqueness of its problems, its size and its age, certainly the textile industry should be the outstanding exponent of the research method. It is not. There is in this country more real research going on in communications, or in academic physics, or probably in such a comparatively minor matter as refrigeration, than in the central field of textiles. For this I believe there are two reasons: first a distrust on the part of many textile men of the research method, and second the fact that the textile industry is made up of a very large number of comparatively small units, no one of which thinks it can afford to pay the price of a comprehensive research program. There is real sanity behind each of these reasons, yet they must be overcome, and the burden of overcoming them is on the research worker and on the textile owner and executive. In our modern tempo, that industry is in danger which is in a static state. Research, with its yield of new and better products, is not a luxury; it is a necessity, often as a defense against encroachment from without. The world in which we live is going to change, politically and sociologically no doubt, but also technically. We have no choice as to whether we will change with it, we simply have a choice as to whether we will change rapidly enough and sanely enough to remain part of the essential scheme of things, or whether we will pass out of the picture. Looms have operated in much the same way since the time of the early Egyptians. Will they always operate the same way? I do not know anything about looms, but I have lived in a modern research world for some time, and I know they [looms] will not [stay the same]. Will we ever make better fibers than nature, in place of what we may call, with due apologies to hard-working organic chemists, our present weak imitations? Well, there are 100,000 or so known organic compounds with all sort[s] of combinations and states of physical aggregation. We have X-rays and polarized light, and ultra-centrifuges with which to study them. There are quite a few thousand men in the world capable of constructive thinking. If we cannot beat a silkworm, I am ashamed of the human race.

Speaking of the silkworm, there is one research procedure which is highly attractive because it is so natural. That is to find out what nature does and why it is done that way. In my complete ignorance of textile matters, I have lately been asking questions about the silkworm. One I have had partially answered, namely, what materials she uses. (Perhaps *she* is not the correct pronoun, but I did not inquire as to that). The list of materials, still partial, is almost as long as the silkworm's fiber and not nearly as simple in configuration. The next question is more difficult. Why does the silkworm use such a complex mixture instead of a relatively simple product? Then too why does she spin a fiber with a triangular section? There might be an explanation in terms of luster, but presumably she doesn't care unduly for appearances. Possibly it is because there is a higher ratio of drying surface to area for a triangular section than for one that is circular, but if so, why doesn't she go the whole distance and make a fluted affair or a ribbon? Why does she make two fibers simultaneously and carefully cement them together? The reason for this may be that if one spinneret gets temporarily plugged up it will not spoil the show, but if that is true, I would like to know how a silkworm clears a spinneret when once it gets plugged without interrupting the spinning process. I'd like to know how a man does it, for that matter. Finally, and this last question involves pure inquisitiveness and nothing else, I'd like to know how a silkworm winds a complete cocoon and finally leaves herself on the inside of it.

To return to the serious aspect of the problem before us, however, let us examine the reasons why there is not more research in textiles. Some will, I know, disagree with me, and contend there is now a great deal more research going on than I have any concept of. Perhaps this is true, but as a simple wearer of clothes, I doubt it. It seems to me that the shape gets out of my suit, when I lounge in it, nearly as rapidly as it used to. It is imperfectly elastic, it takes a permanent set, it stretches visibly in places where I do not want it enlarged. The fuzz, or whatever is the proper technical term, still wears off my sleeves when I rub them on my desk. When it is cold, I still have to put on an unwieldy number of layers of clothes in order to keep reasonably warm. Still, it is only within a few years that the thermal conductivity of textiles has even been scientifically measured. Whether there is a great deal of research underway or not, it appears that there is an opportunity for the profitable employment of more.

There has been a great deal of time and effort wasted in improper or untimely research, a moderate distrust on the part of the layman is entirely warranted with respect to the type of research which consists simply in mixing things up or measuring something without any very clear idea as to the goal and with no correlation with the recorded results of the literature of the subject. The idea is unfortunately prevalent that one may buy a research laboratory ready-made from scientific instrument makers and by hiring a doctorate graduate and turning him loose be securely embarked on a program of research that is bound to yield results that will revolutionize a part of an industry and cause it to produce desirable products with a high margin of profit. The trouble with that scheme is that there is everything present but the program, and the mature and experienced brains to guide it. There is not the slightest question but that well directed and intelligent research pays well in any field in which there are important problems to be solved. There are sufficient examples of this all about us so that the thesis no longer requires demonstration. There need be no distrust of intelligent research. The prime essential to an intelligent research program is not, however, apparatus or funds or young research workers; it is . . . a mature research director with a thorough knowledge of his field, a standing in his profession, a vision of possibilities, a courage to attack the unknown, a patience that is inexhaustible, and a kindly humanity that will cause his co-workers to rally about him with enthusiasm. Find such a man, and the rest of the research laboratory and program will appear. Such men are scarce, and they come high. It is my reasoned opinion, however, that no comprehensive research program can be successful in the long run, except by sheer luck, unless it is centered about such an individual.

There are more of these research directors about than might be suspected offhand. For a life of research is exceedingly attractive to the youngster of inquiring mind; they have simply not as yet been discovered. My advice then to the textile or other executive who's considering the launching of a research project is to seek out a[n Eli] Whitney or a [Frank] Jewett of about age 30, be very sure you have the right man and then keep him entirely happy. The results may come slowly, and the more comprehensive is a research, the slower it will come to fruition, but [results] will come nevertheless and they can be as far-reaching in their effect as is necessary to keep an organization busy for a generation.

The question remains, how is an executive to recognize such an individual when he comes in contact with him? There is the recognition which comes to any man of outstanding technical skill, of course, but it does not go far on the matter of scientific vision and judgement. There is recognition of a man's standing in his profession, his publications, and the opinion of others in his technical field. Unfortunately, these operate fully only when a research man has become established and usually somewhat late in his productive life. I believe that the direct approach to the problem is also needed, that the executive today of an important industry involving a complex technique should himself understand research to a sufficient extent to speak the language and to mix freely and usefully with research workers. It is just as necessary, in my opinion, for the modern executive to understand the examination of fibers under polarized light as to know the mechanics of a spinning frame or be able to read a balance sheet. Of course, he cannot hope to know all the underlying optics any more than he would attempt to operate personally any and all of the machines in his plants or keep up with the intricacies of present-day cost accounting systems; but he can understand those things which are important to his business, and this is now one of them. The executive today should recognize the outstanding research worker because of his own knowledge of the subject matter of research. Incidentally there is no more fascinating study for any man, whether he is directly interested in the results or not, than the record of the work of the researcher worker. Of course, the story of the great inventor is interesting, and the superficial treatments of the popular technical magazines are well [worthwhile] as far as they go. The technical reports themselves, those dry careful records of experiment and conclusion, those painful logical arguments from cause to effect, have a fascination all their own for those who will take the trouble to read and understand them, for they are not light confections. The best of them are solid meat, food for mature intellects, and the concrete expressions of the real material advance of our civilization.

The other reason mentioned why there is not more research in textiles at the present time is the lack of single organizations of sufficient size properly to afford the large expenditures necessary to a comprehensive and coordinated program of research. The nearest approach to such an organization in this country is the United States Institute for Textile Research. Properly supported, it can fill a serious need. Fortunately, there are researchers of all sorts of magnitudes, and individual's problems can be and often are split off for separate attack by relatively small groups. The really inclusive program, however, is very large. It involves a great deal of physical, organic, and inorganic chemistry; the science of colloids; many branches of physics; a great deal of mechanics and optics; X-ray analysis; and several branches of biology. Much is accomplished at present by groups working in these various fields, in private, semi-private, and public laboratories and organizations, exchanging their views and results by means of publication in the various journals. There are single research groups involving workers of some different types, but none in this country involving all types. The large-sized private research laboratory, such as exists in the telephone field, in electrical manufacture, or in the manufacture of certain types of chemical products, does not exist in the textile field, and the results of a single-minded aggressive private attack on the general problems of textiles are hence absent. Its place is taken by various governmental, university, cooperative, and private enterprises which cannot, in their very nature of things, proceed as rapidly or as far. They are doing remarkable work, as some of the recorded results . . . testify, but there is a great deal that remains to be done. Their accomplishments prove that we can learn much and produce useful and desirable things even when the effort is somewhat scattered. There is no field in which there is a greater opportunity for accomplishment than that of textiles. The key to accomplishment is research. It need not be complex, in order to be useful, but it certainly should be intelligent.

3

THE INSCRUTABLE PAST (1933)

In 1933, amid the deepening despair of a Great Depression, Bush was asked by the editors of MIT's Technology Review *magazine to imagine a better future in the distance and then look back from this improved, if imaginary, time to put into broader context the struggles and travails of the day. Alas, Bush is no budding H.G. Wells. The essay falls flat. Part of his impulse for writing, given the context of economic crisis, was to counter the prevailing view that too much technological advance had contributed to unemployment and depression. Then dean of engineering at MIT, Bush advanced the contrarian notion that too little technological innovation, not too much, was what ailed America and the world. As a leader at the premier engineering school in the United States, Bush presumably knew of what he spoke, but he chose a convoluted manner of making his case. Sitting at an unspecified time and place in the future, Bush awkwardly looked back in disapproval at the lack of human techno-scientific capacities. "Take the early 1930s as an example. From this distance" in his imagined future, "the mechanical aspects of that time certainly appear grotesque. . . ."*

The article was Bush's first in a general magazine or newspaper. He would publish many more for everyday readers, communicating more effectively than in this one (where among other lapses, he regrettably addresses his audience as "men"). When he tries to poke fun at himself and his contemporaries for expecting too many novel inventions to arise too quickly, and of the general tendency for people to underestimate the degree of difficulty in launching radical advances, Bush skewers what we now view as the "hype" around emerging technologies. He also anticipates the concept of future shock, articulated in 1970 by Alvin Toffler, a popular writer and influential thinker who argued that technological change can overwhelm people, even leaving them disoriented and stressed. But because he chose an awkward manner of expression, Bush often fails to hit the mark, confusing the reader and obscuring his larger purpose.

The significance of his first essay, however forgettable as an experiment in witty futuristic writing, lies in the judgement of historians of computing who celebrate the first description of a device that Bush would come to call the "memex," an abbreviated term of his own invention that combines the words memory *and* extender, *or* extension. *"By depressing a few keys," a researcher could "instantly" retrieve from thousands of books "a given page," which would be "projected before him" on a screen embedded in his desk.*

Bush also envisions a future where personal libraries, of the sort academics adore and professionals depend on, will have undergone major changes because of technologies of miniaturization. Revealing his broader vision about humanity and our tools, Bush shows he values looking back to "bygone days" in order to gain insights into how innovations occur in the past, and thus may emerge in the future, and he exposes his fundamental pragmatism by expressing the view that utilitarian motives shape innovation, especially the drive for greater ease, comfort, and convenience.

THE INSCRUTABLE PAST

A review of the mode of living of our forefathers [in the 1930s], if it is to be useful, should be sympathetic in its attitude. The lapse of time often obscures the difficulties surrounding a former generation, and we are apt to smile at crudities when a just estimate should rather leave us to marvel that so much was accomplished with so little. . . .

"The Inscrutable Past" first appeared in the January 1933 issue of *Technology Review*, MIT's alumni magazine; in 1946, reprinted in *Endless Horizons*, a long-out-of-print collection of essays by Vannevar Bush (Washington, D.C.: Public Affairs Press, 1946).

We read of the trials of the men of that day and wonder that they could have been apparently content with their mode of life, its discomforts and its annoyances. Instead, we should admire them for having made the best of a hard situation, and treasure the rugged qualities which they exemplified. It is possible that by taking our minds back, divesting them of their modern knowledge, and then studying these bygone days in an attempt to really appreciate their true worth, we should lose some of our satisfaction with respect to the technical accomplishments of our own generation and be better prepared for advance. At least it is worth the attempt. . . .

Lunch for our professor [in the 1930s] was a ceremony of a sort. Nearly all the ingredients were of natural origin with very little in the way of synthetics or products of the biological industries. The foods were attractive in their way, but chosen almost at random and served in circumstances that were somewhat appalling. To the discomforts noted in his office were added in the dining room a complete intermixture of the odors of all the several dishes that a blind man could have told the instant he entered the door of a restaurant. In fact, he would not have had to depend upon his olfactory sense, for the dishes were of various glasses and ceramic materials (as well as in some cases the tables themselves), so that there was plenty of notice from the impact of hard materials.

On his return to the office, the same hectic round would continue. Perhaps in the afternoon the incessant clatter of typewriters would be especially annoying. For letters were written on typewriters, and there was a great deal of letter writing. That was the only way of practically conveying ideas outside of the archaic telephone and personal visit.

The library, to which our professor probably turned, was enormous. Long banks of shelves contained tons of books, and yet it was supposed to be a working library and not a museum. He had to paw over cards, thumb pages and delve by the hour. It was time-wasting and exasperating indeed. Many of us well remember the amazing incredulity which greeted the first presentation of the unabridged dictionary on a square foot of film. The idea that one might have the contents of a thousand volumes located in a couple of cubic feet in a desk, so that by depressing a few keys one could have a given page instantly projected before him, was regarded as the wildest sort of fancy. This hesitation about accepting an idea, the basic soundness of which could have been tested by a little arithmetic, is worthy of more than passing notice. For the tenor of the age was to welcome new inventions and theories. In fact, the man on the street was wont to visualize scientific triumphs as *fait accomplis* even as they were being hatched in the laboratory. He combined a simple credulity on some things, not erased even by the singeing of the Big Bull Market of the late Twenties, with a strange resistance to others. It seemed that the greater the technical difficulties which accomplished some really revolutionary proposition, the more casually the ordinary citizen accepted its consummation as being temporarily delayed but a fortnight or so.

Television was a case in point. To read the contemporary popular accounts one would suppose that the basic problem was solved at least once a month for several years. Yet the public seemed not to mind this crying of wolf, and quietly ignored simple analyses which showed that to transit the image of a man's face in recognizable fashion would require 50 times the amount of communication channel adequate to transmit his voice. And when the progress of television proved to be exceedingly slow (like many other things which in the Thirties were asserted to be just around the corner), the layman was positive that the retardation was because of some corporation's machinations.

Somewhat the ordinary fellow of the Thirties, though he was by no means so witless as he deemed himself when he counted up his stock market losses in the earliest years of the decade, was quite muddled in his thinking process as seen from our present vantage point. He would, as I have said, readily accept the solution of such a complex thing as television to be imminent—as something he might find poking its way into his bedroom unawares on a bright Sunday morning. But he would consider a reasonable improvement in such an elementary thing as the arrangement of sleeping car space (it was really being tried by the railroads at the time) as incapable of realization for a couple of generations at least.

All about him he could see bridges, viaducts, steamships, engines, and so on being built in hitherto unprecedented sizes. And, if some publicity agent issued an optimistic statement to the press that in the coming year they would be built twice as large again, he'd accept such a radical prediction with little emotion. Yet, when it was proposed to make it practicable for those who were neither too fat nor abnormally tall to undress in an upper [train] berth, his reaction was likely to be expressed in the quaint vernacular of the day by some such expletive as "baloney" which, it seems, signified intense credulity and impatient skepticism. . . .

4

THE WARREN WEAVER LETTERS ON THE FUTURE OF COMPUTING MACHINERY (1933)

These three letters to Warren Weaver, a mathematician and a program officer for an important research funder in the 1930s, illustrate how Bush mobilized resources and focused attention on his own innovative research in electro-mechanical computing. Even as Bush worked as a research leader and administrator, first at MIT and later in Washington, D.C., Bush stayed at the forefront of innovations in computing through a series of machines he designed and built in Cambridge, Massachusetts, with the financial support and encouragement of Weaver and the Rockefeller Foundation. Bush also helped other scientists, notably in Britain and Norway, to duplicate his machines and to even improve on them. The culmination of Bush's efforts was the so-called Rockefeller Differential Analyzer (RDA), which the historian of information Robin Boast has described in his book The Machine in the Ghost *(2017) as "one of the most important calculating machines of the Second World War."*

Working with his graduate students Herbert Stewart and Harold Hazen, Bush developed a general-purpose analogue computer that in 1931, he described for the first time in a scientific paper in the Journal of the Franklin Institute, as a "differential analyzer." The machine was an analogue computer that could solve differential equations with as many as eighteen independent variables. Or as Bush recalled in his memoir Pieces of the Action, *the machine could solve "tough equations." As historian Boast recounts, "mechanical analyzers had to be set up by hand, usually with spanners (wrenches) over many hours and days. . . . In addition, any wear to parts, gears or integrators, could add unacceptable errors to the system. Bush overcame many of these problems with the RDA by replacing many of the mechanical devices with electronic components. Vacuum tubes, relays, sensors, and amplifiers were all now controlled by a program punched onto paper tape. This vastly reduced the error and increased the accuracy and speed of the RDA by a factor of ten over earlier differential analyzers. For this reason, the RDA, and its successor the RDA2, continued in service throughout the war and well into the 1950s." Boast adds, "One of the most famous uses of the RDA2 was for calculating, in reverse, the trajectories of the German V2 rockets at the end of the war. This allowed the Allies to discover and ultimately destroy the launching sites of these first ballistic missiles."*

In Bush's letters to Weaver, we see the origins and antecedents of this landmark achievement in early computing.

"I HAVE BEEN DREAMING IN A RATHER DEFINITE WAY" (JANUARY 6, 1933)

Dear Dr. Weaver:

In view of our conversation when you were last at [MIT] it occurs to me that you might be interested to know somewhat more in detail what I am doing in regard to plans for an improved Differential Analyzer. While the present machine is proving very useful, and while we are concentrating on making it perform, I have been dreaming in a rather definite way of the next step in this general program. The success which we have had with some of the difficult equations of mathematical physics has led me to believe that we can go quite a long distance in this direction and obtain fruitful results for our efforts. Hence while I realize fully that it will take some time before it will be desirable to undertake a new program of construction, I have been making plans so that we might see our major difficulties and give them thought.

Vannevar Bush to Weaver, Rockefeller Archive Center (Sleepy Hollow, NY).

The scheme on which I have been working is one that I have used before. I first draw up a statement of what I should like to have a new machine accomplish. Then, with this [objective] before me, I write a set of specifications, making it as detailed as possible in order to bring out the major points of difficulty. These [major points], with several discussions with some of my associates, I have been revising from time to time. I think that you may be interested to look at the specifications in their present state, realizing that they are still crude and that they are constructed primarily as a means of focusing ideas. Accordingly, I enclose a set, and if you have time to glance them over, I will be very appreciative of your comments. Incidentally, also I am enclosing a photostat of the Atlantic City exhibit of the differential analyzer, which I thought you might care to have.

Sometime, when you are going to be in this vicinity, I would highly value an opportunity to discuss this general program with you, for while I realize that it is a long-time program in every sense, I am anxious to direct it in such a way that it will ultimately produce real results of value not only in the field of engineering (in which I am primarily interested), but also as a working tool for the scientist.

Yours very truly,

V. Bush

"TO CRYSTALIZE IDEAS SOMEWHAT MORE"
(APRIL 15, 1933)

Dear Dr. Weaver:

. . . In order to crystalize ideas somewhat more in regard to the advisable extent of the next [Differential Analyzer] model. . . . I have had time to pore over the specifications which I sent you recently and bring them into somewhat better shape, and also to produce an appendix, a copy of which I enclose. The more I think of this matter, the more I become convinced that it is going to become possible to produce something quite startling. This last appendix contemplates a practically instantaneous change from one problem to another or from one set of starting conditions to another. It would make the net speed of the device very large.

A lot of complexity is involved, and a good many relays, but this same thing is true of such an affair as the automatic telephone exchange, which is a pretty reliable thing.

With pressure of affairs at present, and particularly of budgetary matters, I have not had too much time to ponder on this affair, but I have some hope that the coming summer may be sufficiently clear so that I will have time to really get involved in bringing this into shape. Certainly we are overcoming some of the major difficulties even now . . . in a new differential analyzer development.

I plan to keep you informed from time to time concerning the progress that is being made and the plans for the future, and I would be very happy to talk them over with you at length sometime. In spite of all the obstacles, I have gotten to the point now where I am convinced that we could build an advanced type of differential analyzer which would make the handling of ordinary differential equations a much more satisfactory thing than it is today.

Very truly yours,

V. Bush

"CONTEMPLATING SOMETHING OF A RADICAL DEVELOPMENT" (JULY 7, 1933)

Dear Dr. Weaver:

Dr. [Svein] Rosseland [a young astrophysicist from Norway who visited Bush and his lab] took leave of us a short time ago, and is returning with sketches, photographs and the results of two problems which he and [Samuel] Caldwell worked out together on the Differential Analyzer. I feel sure that [Rosseland] is in a position to construct at reasonable cost a machine which will be essentially a duplicate of ours except that it will have improvements which will render it somewhat more flexible and reliable as a result of our experience. Rosseland left with a great deal of enthusiasm, and I feel [he] is in a position to make the most of this form of attack on his problems.

He has in fact the thought of going much further with the Differential Analyzer than the mere construction of an improved model along the present lines and is contemplating something of a radical development. Sometime before very long I think that you and I should discuss this entire matter of the further development of this type of machine. I believe we can now start with the premise that an intense development is justifiable, as we seem to have in our hands the key to quite a number of situations where laborious analysis makes rapid progress impossible.

My own feeling is that the best possible way to proceed would be to help in the construction of one or two machines along the present lines with improvements of routine nature, and simultaneously to concentrate on one radical development. With all modesty I think that this radical development should be in my own hands, for now I have a background of experience which is rather necessary in order to appreciate the difficulties involved in the radical step forward. It is rather hard for anyone without this experience to appreciate the difficulties involved in a radical step forward. It is rather hard for anyone without this experience to appreciate the extent of the struggle which is necessary to overcome even simple mechanical difficulties in a device of this sort. I do not wish to object to a radical development on the part of anyone. In fact, the most desirable thing would of course be for several people simultaneously to be attempting the larger step, but such a duplication of effort would be somewhat expensive. The best all-around procedure hence seems to be one to pioneer in a field which is expensive even with the best of economy. . . .

Dr. [Douglas R.] Hartree [the British theoretical physicist] is now [visiting MIT] with us . . . Rosseland very kindly turned over some of his sketches and photographs to [British scientists] and this will shorten their labor considerably. Hartree has put a problem on the machine, working through all the details himself, and I feel that this is by far the best way for him to learn the idiosyncrasies of the device. Of course the work will go slowly as the main object will be to bring Hartree in contact with various aspects of the use of the machine, but nevertheless I believe that he and [Samuel] Caldwell [a new PhD who Bush advised and who's dissertation was devoted to the Differential Analyzer] will produce results of value during this month.

Hartree has developed much enthusiasm as he has realized the power of the device, and his joy at watching the performance has been well worth the effort we have made to render his visit profitable to him. My own enthusiasm is not waning a bit. This present model has just taken us over the brow of the hill, and from now on we can see the results of our labors. I feel sure that the new development will be far in advance of the present, and that it will change labor at a snail's pace into rapid progress in pure and applied physics, as well as in such rather remote fields as the very important subject of econometrics.

I believe you are going to return to this country sometime during this summer, and I hope that we can get together. I have written this letter rather frankly, and I would like to feel that you are thinking the matter over with my point of view in mind before we have a chance to discuss the situation.

Very truly yours,

V. Bush

5

THE PERSISTENT FALLACY OF THE
ABSENT-MINDED PROFESSOR (1933)

Bush passionately believed that scholars, engineers, and scientists have an obligation to speak clearly and address issues of public importance. In this essay in honor of Elihu Thomson, the English-born engineer and inventor (1853–1937), Bush seeks to both celebrate Thomson and promote a new attitude toward academic research. Universities, as much as other professionals, can make a positive difference in the world, Bush insists. With Thomson himself in the audience, Bush affectionately declares, "You have showed us that a man may be truly a professor and at the same time very practical. And one thing more I wish to emphasize. You have shown us that a scientist or engineer may be, even in this complex world, versatile and yet not superficial." With these words, Bush could be describing . . . himself.

THE PERSISTENT FALLACY OF THE ABSENT-MINDED PROFESSOR

There has long been, on the part of the layman, an inclination to associate the title "professor" with the pedantic and the cloistered. To fully

---

Excerpted from "In Honor of Professor Elihu Thomson," text of an address by Bush to mark Thomson's eightieth birthday, March 1933. The full text can be found in *Science* magazine, May 5, 1933.

earn the title, it is sometimes popularly supposed that one must cultivate an intense absent-mindedness and a lack of knowledge of the everyday affairs of the world. You have certainly done much to dissipate this idea, and your colleagues rise and call you blessed. That there exist professors who can put up a window shade while conversing in ordinary language, or who can set the timing of an automobile as a form of recreation, is a fact of the modern world which is still incredible to the laity. That there is one who has hundreds of patents to his name for all sorts of things which very practical men have been very glad to use has given the title a significance which is well added to its dignity. I hope that every farmer who uses a cream separator will realize that it was invented by a professor, for it will do much to counter that persistent fallacy that those who think deeply are necessarily beings apart.

The world progresses only so fast as it learns to pass on its accomplishments from one generation to the next. The material things, those that are of direct utility, machines and structures, can hardly be passed on at all today, for obsolescence is much more rapid than deterioration. The written word, the recorded science of yesterday, is here today and we may delve and explore for the thoughts of the pioneer in order that the gains made by those who blaze the path may not be lost to those who follow. But there is a means of transfer of accomplishment which transcends either of these, just as the speed of thought is greater than the working of tools or the reading of words. It is the direct contact between the alert mind of youth and the seasoned and inspiring mind of experience. This is education in its essence. All the paraphernalia of the college, the faculties and curricula, the tall towers and the basement laboratories, have but one purpose, to bring together in an atmosphere of progress the youth and his teacher, and to make it possible for them to work.

It is not by chance that you—who have been scientist, inventor, organizer, engineer [and] man of business—have been for sixty-odd years [a] professor. It is because that profession holds for you, as it does for us, your colleagues, the key to an accomplishment that is more truly satisfying even than the pushing back of the boundaries of science or the application of the fruits thereof to the material benefits of man. There must have been a glorious feeling of triumph when you first watched an electric arc stream out and attenuate in a magnetic field. The satisfaction of electrically making one piece of metal where there were two before was intensified by the realization of what the process would mean in the comforts of life to all. But I venture the thought that these were not as stimulating or as lasting as the satisfaction which was yours when you first caught in the eyes of a youngster that gleam of fire which told you that a spark of your own wisdom had transferred to another mind. The joy of teaching is deep seated in our primary instincts, and there is [in teaching] an immortality more specific than fame.

You have taught us many things of less tangible nature than how to make a transformer yield constant current. You have showed us that a man may be truly a professor and at the same time very practical. And one thing more which I wish to emphasize. You have shown us that a scientist or engineer may be, even in this complex world, versatile and yet not superficial. I recall that you were first a professor of chemistry and of mechanics, and you are now professor of applied electricity, and you have been organizer, inventor, man of business, engineer, astronomer, executive, [and] philosopher. In these days, when there is a tendency to specialize so closely, it is well for us to be reminded that the possibilities of being at once broad and deep did not pass with Leonardo da Vinci or even with Benjamin Franklin. Men of our profession—we teachers—are bound to be impressed with the tendency of youths of strikingly capable minds to become interested in one small corner of science and uninterested in the rest of the world. We can pass by those who, through mental laziness, prefer to be superficially interested in everything. But it is unfortunate when a brilliant and creative mind insists upon living in a modern monastic cell. We feel the results of this tendency keenly, as we find men of affairs wholly untouched by the culture of modern science, and scientists without the leavening of the humanities. One most unfortunate product is the type of engineer who does not realize that, in order to apply the fruits of science for the benefit of mankind, he must not only grasp the principles of science but must also know the needs and aspirations, the possibilities and frailties, of those whom he would serve. There are students who may realize this fallacy only when it is too late.

6

STIMULATION OF NEW PRODUCTS AND NEW
INDUSTRIES BY THE DEPRESSION (1934)

The global economic depression in the 1930s spawned new challenges in thinking about technological advance and the capacity of markets to deliver steady benefits to democratic societies. Writing "The Stimulation of New Products and New Industries by the Depression" in late 1934, when the United States remained in the grip of the worst economic crisis in its history, represents an important, indeed singular, statement of Bush's mature views on innovation, capitalism, and the relationship of science and engineering to both. These pithy ideas, which were written bluntly as talking points for a speech to a meeting of industrial engineers, would inform and undergird Bush's practices and principles for decades to come. While Bush expresses some outmoded and old-fashioned attitudes about economic affairs in his emphasis on the value of new products and new industries, there are intimations of the Austrian economist Joseph Schumpeter's concept of "creative destruction"—the idea that technological advance helped as well as harmed and created winners as well as losers. The concept, which Schumpeter popularized in English in 1942 with the publication of his book Capitalism, Socialism, and Democracy, *matched the experiences of Bush as an educator, researcher, and commercial innovator and jibed with Bush's own view that the severe pain of the Depression could incubate new forms of technological advance, or as he put it, "resets the stage for creative enterprise." The result of the process can be "immediate disorganization," but an "ultimate increase in standards of living." Bush's concern with the business and economics of science set him apart from many contemporaries and anticipated the emphasis that corporations and investors began to give science and research later in the twentieth century, and the importance to innovation of "men of vision," or product champions who, as Bush writes, "bring these" disparate strands of the process "together." In this speech to an association of engineers, Bush prepared his thoughts on the essence of innovation in a list, which he preserved, and the list is presented below as found in his notes for the lecture.*

STIMULATION OF NEW PRODUCTS AND NEW INDUSTRIES BY THE DEPRESSION

1. Depression is the fever by which civilization rids itself of infection.

    Old and new ideas of fever in disease.
       Attempt to cure fever replaced by attempt to remove cause.
       Artificial fever. Malaria. Inevitable result of human nature.
    Danger that the fever will kill.
       Recovery inevitable if it does not.
       A time of nostrums and isms. No especial "ism" in mind.
       Need of rest for recovery.

2. Depression produces new things.

    Dams the outlet while men's minds work.
       More thinking than in 1929
       Resultant surge when dam breaks.
       Breaks conservatism, not only in political arena, but in business.
       A time of change. Resets the stage for creative enterprise.

3. Before we consider how in detail, let us examine whether new things are desirable

    Immediate and ultimate impacts
    Impact of new developments

"Notes for Talk Before Society of Industrial Engineers," Cambridge, Mass., November 22, 1934, Archives of the Carnegie Institution of Washington.

Trucks and Railroads
Synthetic Indigo
Radio and Pianos
Immediate disorganization—ultimate increase in standards of living
    Electrical industry
    Life of semi-skilled today and century ago
    Conquest of Disease
Results not all desirable.
    Transportation, communication and a shrinking world
        Can men live closer together in peace?
    Use in war
    Public and government problems become more complex
    Increased education. But mass response is still erratic.
        Swayed by propaganda, advertising.
    Complicated problems before an emotionally unstable citizenship.
    Decline of democracy. A blind mass rushes on.
    But who would return to a world with disease rampant?

4.   Much remains to be done

Even in minor ways
    Crude automobiles, crude trains (bumps, noise, delay), crude houses (thermal heating, noises)
In major ways
    Economic science. Slow progress toward security for the individual.
    The money muddle. Hole in the ground.
    Cost of mentally feeble
    Agricultural products—fibers, fuels, plastics from the farm

5.   What is essential to new product?
  a. Sound growing body of underlying science

       America taking its place.
       Little grasped even today. A few torchbearers

  b. Ingenuity and resourcefulness

       America never lacks inventors. But it usually misunderstands and misuses them.

c. Funds for development.

Which are plenty even now. The vital need for opportunity to make an unusual profit.

The tax system. Russia can catch up, but can she forge ahead.

d. A market: the American public is all too avid for new things.
e. Sound engineering: to economically bring science to function in useful ways.
f. Industrial conditions permitting manufacture at reasonable cost.
g. Someone to bring it together. This is where I come to this audience.

6. Paint two pictures of the businessman in this situation:
   a. He doesn't know the difference between science and philosophy. He thinks a mathematician can add a column of figures. He believes there are only [twelve] people who understand Einstein. He thinks television was invented a few years ago. He believes Watt invented the steam engine and that Marconi originated radio waves.
   b. He believes all inventors have long hair and should be shut up in cubicles. He thinks they should have a knowledge of business, and that if they do not they should be instructed in the same way that instruction is given in the art of poker.
   c. He thinks there are two kinds of engineers, those who pull throttles and those who push stopwatches.
   d. He believes the way to sell any product is to repeat the same inane slogan millions of times and paint store fronts in repulsive color combinations.
   e. He thinks of the businessman as a super-being apart, who has the only true evaluation of success in the world, to whom homage should be paid.

This [image] is extreme and overdrawn.

The other picture is this: There have been and there will be great technical advances in this country. They will be brought about as they have always been by men of vision. Men who realize they play but one part in the drama and who seek to understand and work with the other players. In the realization that fruits of progress are many, and profit only one. That the satisfaction of creation, the joy to ministering to the happiness of one's fellows, transcend profits. That these satisfactions are sought by hard-headed, competent men.

7. Success in the promotion of new products in the past has often gone to the opportunist who chanced upon and exploited a casual advance.

The cream is skimmed. The success in the future will go to that business man, one among many, who evaluates and utilizes all the factors essential to real progress in this kind of world.

To him the path is open, if sanity returns, and the fever passes.

7

THE BUSINESSMAN IN THIS SITUATION (1934)

Bush had a strong interest in business, which he saw as a potential partner of engineering and logical patron for the activity of research discovery. While acknowledging that executives and business owners could be crassly concerned with profit and blind to culture, he also sought out exceptions among the pecuniary tribe. During the Depression, when both business and technology drew blame for joblessness and the economic crisis, Bush advocated that business begin to take more progressive approaches toward scientists and engineers, and to place greater value on knowledge and invention. In this excerpt from notes for a speech he gave at a Masonic Lodge in Lowell, Massachusetts, Bush shows flashes of his irreverence and wry sense of humor; here he at times pokes fun at the very businessmen that he professes can also sometimes possess signal "vision." The core of his argument lies in the recognition that business can do more to work with and value engineering, science, and innovation. If such collaboration between the world of commerce and technology strengthens, he insists, an optimistic view of the future is justified, so long as Americans "understand the need for stability." In short, Bush held out hope that the Depression would pass and that innovations would once more propel gains in material life for the many. Given the role of business in war and postwar America, especially in regard to the rise of a military-industrial complex, Bush's formative views on "the businessman" are significant. He organized his ideas in a list, which he preserved, and the list is presented below as found in his personal papers.

THE BUSINESSMAN IN THIS SITUATION

An extreme and overdrawn statement [about the businessman]

1. He doesn't know the difference between science and philosophy. He thinks a mathematician can add a column of figures. He believes there are only 12 people who can understand Einstein. He thinks television was invented a few years ago. He believes Watt invented the steam engine and that Marconi originated radio waves.
2. He believes all inventors have long hair and should be shut up in cubicles. He thinks they should have a knowledge of business, and that if they do not they should be instructed in the same way that instruction is given in the art of poker.
3. He thinks there are two kinds of engineers, those who pull throttles and those who push stopwatches.
4. He believes the way to sell any product is to repeat the same inane slogan millions of times, and paint storefronts in repulsive color combinations.
5. He thinks of the businessman as a super-being apart, who has the only true evaluation of success in the world, to whom homage should be paid.

The other side, the other picture, is this:

1. There have been and there will be great technical advances in this country. They will be brought about as they have always been by men of vision. Men who realize they play but one part in the drama, and who seek to understand and work with other players. In the realization that fruits of progress are many, and profit only one. That the satisfaction of creation, the joy of ministering to the happiness of one's fellows, transcends profits. That these satisfactions are sought by hard-headed, competent men.
2. Success in the promotion of new products in the past has often gone to the opportunist who chanced upon and exploited the casual advance. The cream is skimmed.

Notes for a talk at the Kilwinning Masonic Lodge in Lowell, Mass., November 30, 1934, Archives of the Carnegie Institution of Washington.

3.   The success in the future will go to that businessman, one among many, who evaluates and utilizes all the factors essential to real progress in this kind of world. To him the path is open, if sanity returns, and the fever [of economic crisis] passes.

But overall is the need for stability:

1.   We need have no fear for the future if this is secured. [Yet] I would not [wish to] engender a smug content with things as they have been. The world from its beginning has been tortured by inequities, injustices, and crushing poverty. But the standards of living, and standards of common morality, are rising, slowly and steadily, and the world is a far more pleasant place in which to live and work than it was when the first farm was carved out of these wooded hills. May we live to see the complete disappearance of man's inhumanity to man.

2.   But one does not spur the horse he mounts for a day's journey; nor is a ten-second pace in the first hundred yards the way to begin a mile race. It is because I feel the necessity for real progress so keenly that I see the insistent need for steadying hands in this time of cure-alls and nostrums.

3.   And who shall apply the steadying hand? He who has learned to be his own master. He who has drawn in with the free air of the hills, far from the tumult of the cities, something of the permanence of their foundations. He who has learned to think in terms of generations. He who has courage in adversity.

8

AGAINST ISOLATION AND FOR APPLYING
SCIENCE TO WAR (1935)

Bush's arresting message presages his own future: engineers and technologists can and should work on new weapons and defense systems when their nation's interest demands it. "Splendid isolation," however attractive, he insists, makes peace less likely, not more. Writing at a time when many in Europe and the United States wished to avoid another "Great War," Bush's views were contested. In the world of science and engineering, the legacy of bringing chemical weapons onto the battlefield in World War I continued to erode public trust in scientists. Moreover, the reputations of those scientists who, in the 1930s, remained directly involved in weapons development suffered. Bush insisted that isolationism should be replaced by a more realistic and flexible perspective, which allowed for the possibility that durable peace can be achieved through techno-scientific strength and weapons innovations, especially defensive measures that would counter the growing threat of attacks by air. "Perhaps the worker on antiaircraft is more effectively a worker for peace than his brother who condemns him," Bush writes. "I do not make the assertion, I pose the question."

AGAINST ISOLATION AND FOR APPLYING
SCIENCE TO WAR

Take the exceedingly important problem of international peace. Everyone who is not an enemy of society is a pacifist in the true sense that he would banish war from the world. Every thinking individual with altruism in his soul would labor unremittingly to bring permanent peace among nations. Yet how can we discuss this subject coldly and logically? Its very intensity banishes open-mindedness. The depth of yearning for a solution grasps at straws as would a drowning man. The very earnest individual would personally renounce any bearing of arms under any circumstances whatsoever. Yet the same one will sometimes clamor for our nation to join in the application of sanctions, even to shutting off of necessities from a stated country, which is itself an act of war and may lead to other acts. Certainly if this country were ever faced with the necessity for cold logical reasoning in regard to its international policies, it is at this moment.

Yet those who would find our best contribution in a splendid isolation cry out emotionally against those who would put their trust in collective action to restrain the aggressor. And those who would have us play an active part in policing of the world rail at those who, with equally sincere conviction, would reduce our forces to impotency. Worse yet, many otherwise logical individuals, overcome by the sheer weight of the problem, refuse to attempt to think it out at all, but cry "peace, peace," when there is no peace, and pitifully hope by their cries to still the storm of human conflict and ambition.

Let us consider one aspect of the matter, and pose an unusual question, to see whether logic will carry up surely in its examination. Is a scientist who is busily at work on the perfection of an instrument of war in time of peace ever justified in doing so? Suppose he is developing a new means to surely and quickly locate and bring down an aircraft in flight. What would be the result if he were very successful?

The nations of Europe today, and perhaps of the world tomorrow, are on edge and nervous because of the possibility of sudden invasion by air. The complete removal of this threat would still many a passionate rush for

---

Text of Bush address to Phi Beta Kappa society of Tufts University, November 1, 1935, Archives of the Carnegie Institution of Washington.

armament. The last war demonstrated the practical impossibility of invasion by land of a civilized and reasonably prepared country. It taught the futility of sending battleships against fixed fortifications. Only the path by air remains. If this were removed, a nation could become secure against invasion by reasonable development of a self-sufficient economy and the provision of border defenses. Suppose it were an accepted fact that no civilized nation could be ruthlessly and suddenly invaded? Suppose France were relieved of its fear of invasion from the North, or Germany [relieved of its fear of invasion] from the East? Would the ponderous machinery for the maintenance of peace not have a chance to function? It was the possibility, nay the conviction of the necessity, for mobilization and movement in a matter of days or hours which in 1914 precipitated a war upon the world before it had a chance to think. Perhaps the worker on antiaircraft is more effectively a worker for peace than his brother who condemns him. I do not make the assertion, I pose the question. Yet when I pose it in conversation, I meet the answer many times that *all* development of war engines should cease.

Certainly it should never have begun, but this is wishful thinking; for the application of science to warfare will not cease in the world as it is now divided and governed. . . .

9

THE ENGINEER AND HIS RELATION TO GOVERNMENT (1937)

In this essay, his most significant expression prior to World War II on the role of trained engineers in effective governance and evidence-based political policies, Bush insists, "To be an engineer in these days is to bear a proud title." He writes against the backdrop of what he describes as "the growing technical complexity of our lives." He eloquently describes the accelerating changes that have turned the human adventure into "a crescendo of tumult toward the unknown," driven by the "scientific method." The effect of rapid change is to highlight the need for what Bush wryly terms "the steering gear," which he argues will involve integrating engineering expertise into the core of government and public service. Bush's vision of a society where technocrats are essential appears at the height of the Depression and years before the start of World War II. In closing, Bush introduces a theme he will harvest in years ahead; he insists that while "geographical frontiers have disappeared . . . the frontiers of science and technology still remain." So humans "can still build trails in the technological advance." In so saying, Bush anticipates an argument he will make more forcefully and definitively in his 1945 report to the President of the United States, in which he will declare that the frontiers of science and technology are "endless." Far more than in his 1945 report, Science, the Endless Frontier, *here Bush seeks to balance technical and scientific expertise with democratic participation in a system of representative government. The task of identifying and maintaining this balance remains as daunting today for engineers and engaged citizens as when Bush wrote these words more than eighty years ago.*

THE ENGINEER AND HIS RELATION TO GOVERNMENT

Whatever one may conceive to be the future of civilization, whether one is pessimistic or optimistic in regard to the long pull, one thing stands out—we are certainly on our way. The world is full of change. Our systems of government, not only in Europe, but here in America, are in a state of flux. We live faster, occasionally longer, and always in a new pattern of relationships between man and man. We are headed somewhere in a hurry, whether that somewhere be a grand crash, a reversion to a long sleep, or a more abundant life.

The scientist and the engineer started us on this path. They wakened us from the semi-peaceful somnolence of a century ago, shrunk the world to a fraction of its former size, and placed the well-oiled wheels of the scientific method under us; and we roar forward in a crescendo of tumult toward the unknown. Now it is suggested that some attention should be paid to the steering gear.

Much attention has been paid to equipping and speeding up the conveyance in which civilization rides, and comparatively little devoted to safety devices. The medical profession has conquered one plague after another, engineers and health officers have extended sanitary measures, and the mentally defective benefit along with everyone else, breed rapidly, and may inherit the earth. Bringing science to agriculture makes possible the production of food by fewer men on less acreage, and our farmers have no place to go. Individual transportation is cheapened, and automobile accidents kill 30,000 people a year. The airplane applies a multiplying factor to the useful lives of busy and creative men, and brings hostile nations within grasp of one another's throats.

Since the dawn of history, the acquisition of skill and knowledge has rendered more complex the relationships between man and man. It is the function of government to control these relationships. Today the burdens and responsibilities thrown upon government are increasing by leaps and bounds, and government everywhere is staggering under

---

Excerpted from the text of an address Bush delivered to an engineering convention in Milwaukee, Wisconsin, on June 22, 1937. His text was reprinted in its entirety in the journal *Electrical Engineer* in August 1937 (stored as part of the Transactions of the American Institute of Electrical Engineers). The July 1937 issue of *Science* magazine published the concluding portion of this speech.

the load. The kaleidoscopic changes we have witnessed in governments of the nations are a direct result of the growing technical complexity of our lives. Every advance increases the rate of advance, in accordance with the law of acceleration of Henry Adams. Whether democracy in America can successfully carry the modern burden and the burden to come, and endure, remains to be demonstrated. Certainly this depends in no small measure upon the attitude of the great professional class in general, and in particular upon the attitude and effectiveness of the engineer in his relation to government. It is this relationship that is examined in this address. . . .

The engineer, who applies science in an economic manner for the benefit of mankind, certainly has as close common interests with the social scientist as with the physical scientist. Yet the organizations of the social scientists are not greatly participated in by engineers.

Since the beginning, the engineer has cooperated closely with the physical scientist; and in fact many engineers are themselves physical scientists of no mean accomplishment. We have reached the point where engineers are learning to work in similar co-operation with social scientists, but the relationship must proceed much further than this. It is necessary, in connection with many of the problems that come to him, that the engineer be himself a social scientist—not a dilettante, but a deep student and scholar in his own right in various branches of the social sciences. Only thus may he be properly cognizant of the source from which his activity derives, and also properly cognizant of the social conditions under which it becomes applied. Long familiar with applied economics, he now finds pressing upon him other branches of the broad science of human relationships. The colleges have a duty to perform here, and so have the professional engineering societies. Certainly their meetings and publications should generously reflect this growing need and interest.

. . . To summarize this matter of organization: Real attention is being given to the professional advancement and the technical interests of the profession. We have an enormously complex system of organization of scientists and engineers in this country, and yet no effective single central organization representing *all* engineers and expressing their viewpoints on public questions. We have an elaborate mechanism for bringing advice to bear on scientific and engineering problems as they arise in government, and this mechanism is not utilized to the full.

What is to be done about it? . . . The technique of applying the pressure of engineering opinion on great public questions is only one aspect of our problem. Another aspect involves the advice by engineers to government on specific technical problems. This is a large question, and one that involves many of us in one way or another, as citizens and taxpayers as well as engineers. . . .

One cannot give sound advice on important engineering matters without spending considerable time and money. This is the function of the independent consulting engineer. We will not be on sound ground in this country until government, on a basis of adequate and dignified fees, calls for the opinions of independent consulting engineers whenever it has an important engineering problem. This [government] does not do at the present time to any determining extent. If, when the subject has been deeply studied and reports have been presented, the government wishes review by distinguished boards, it will always find men ready to give their services as a matter of public duty. The main reliance, however, must be upon independent consulting engineers, and I wish to make a plea on their behalf.

There are many engineers—many able engineers—in government itself, and these are utilized by government when it has an engineering project to carry out. Army engineers have carried forward on a high plane many outstanding engineering works. . . . But the government engineer is not an independent engineer, and the latter is sorely needed. Given a definite project the government engineer can carry it forward; but he cannot at the same time say that it is a foolish thing to carry out at all, even if his engineering studies convince him that it is. Here is a point at which a democracy is at an advantage compared with an absolute government. The dictator has *only* government engineers—units in a rigid machine. Independence of thought and speech there cannot be tolerated. Yet having the advantage as a democracy of the presence of engineers of real independence, we do not make use of them. This is partly because truly independent engineers are becoming rare; partly because unfortunately government is sometimes not anxious that the full truth be known; partly the fault of the engineers themselves. This matter is worth discussing briefly, for it is truly unfortunate if one of the great assets of democracy is being thrown away.

The rise of great industries in this country, with their own engineering organizations, has restricted the field of operation of independent consultants. The tendency to extend free engineering services as part of the sales programs of large companies, similarly has encroached. Fortunately industry by and large cannot maintain engineering departments capable of coping with the unusual, and these peaks are surmounted by calling in the temporary services of independent engineering organizations. Yet the way of the consultant has not been easy, and the number of men who are truly independent, who have seasoned opinions based on wide experience, is not large. There should be more utilization of men of the type of John F. Stevens called for service at the Panama Canal [on which Stevens served as chief engineer from 1905–1907]. If it were our practice in this country for government to employ independent engineers frequently, the number of such engineers would be greater. When government calls on the engineer at all, it usually attempts to do so on a niggardly basis. It appears to attempt to starve out a group upon which it distinctly needs to lean.

But part of the fault lies with engineers themselves. While we deplore any reluctance on the part of government to let the full light of reason play on its plans for engineering works, we must admit at the same time that the approach of engineers often has not been based upon a sufficiently broad consideration of these very matters. To show that a government engineering work will not pay an adequate financial return on the original investment is not necessarily sufficient to condemn it; yet engineers are prone to limit their considerations to a strict cost and yield basis. The building of a battleship cannot be justified on this basis. The setting aside of a national forest should not be thus approached with limited logic.

Do not think that I advocate letting down the bars of strict reasoning to which all engineering works should be subjected. I have no sympathy with any waste of public money. To build a great dam to supply electric power in a region already amply supplied with power, to irrigate land in a region of no inhabitants while farmland stands idle close by, to render navigable a stream that proceeds into a wilderness, are fool pieces of work in any language. Yet I would have the engineer join with the economist, the sociologist, the student of government, that he may grasp problems in their entirety. . . .

Another important way in which the engineer makes contact with the government is in connection with the legal system, both in law enforcement and in the administration of justice in the courts. . . . There is a real need for close association of scientists and engineers with the legal system at many points, especially in the patent system. The reason is clear. The determination of any legal question depends jointly upon the law and the facts. In a modern technical world, the facts are beyond the comprehension of the layman. When dealing with a scientific or engineering subject, the most eminent jurist or attorney is usually decidedly a layman. The result is often sad. Decisions are rendered by judges to whom the facts of a case are in essence incomprehensible. Present procedure is expensive, indeterminate, and sometimes ludicrous. Details of the procedure aside for the moment, the real reason for this situation is the unwillingness of the legal profession to admit to a basis of partnership the scientist who understands the technical facts of modern civilization, with the attorney who understands the law. We have the spectacle of opposing experts, cross-examined by lawyers who have a week's cramming as a background in the subject under consideration, for the benefit of a judge whose scientific training ended at "Physics 1." The childlike faith of most attorneys in this process of elucidating technical facts is beyond comprehension. To the technical man on the sidelines, it is often evident that the discussion proceeds to about page 20 of an elementary text, when the true answer lies on page 500 of an advanced treatise. The general atmosphere, charged with suspicion, progressing at a snail's pace, is such that the majority of scientific men engage in legal matters just as little as possible. To expect men of great scientific attainment generally to be willing to take part in this procedure is expecting a great deal from the human race. Yet the members of the legal profession generally regard the presence of a technical adviser to the court, not subject to cross-examination, as an anachronism, and they are perfectly sincere and honest in the opinion. The dilemma is clear. The legal profession, which controls the system, cannot itself or through its artifices deal justly in the type of world in which we now live. It will not have the true co-operation of the best scientific and engineering minds in expeditiously arriving at justice until it welcomes them to something besides a subordinate status. In some of its phases, the legal system has been dangerously close to breakdown, and no small portion of this situation is the extent to which it is bogged down in a scientific morass. If breakdown comes, it will be the fault of the profession that molds its affairs and determines its form. Scientists and engineers stand always ready to aid in a matter of public concern and on a basis of professional partnership. In this connection the independent consulting engineer can be of real service in many ways, but space does not permit a detailed examination of them.

Throughout this address I have emphasized the value of independence. So long as this is maintained and there is effective guidance of affairs by an independent professional class, I have no fear for the future. A true democracy, given this support, can compete with dictatorship and prevail.

It was independence of thought, freedom of action, the opportunity of a vast untamed domain that built this country and gave it the highest standard of living in the world. The geographical frontiers have disappeared, but the frontiers of science and technology still remain. Those qualities which built a trail into the wilderness can still build trails in the technological advance. The same qualities of courage, resourcefulness, and independence, which opened the nation, are as necessary today as ever. . . .

The free operation of professional classes, motivated by public zeal and altruism, is an anchor upon which our democracy depends to hold it through the storm. There is a great obligation upon the professional man to speak clearly, to insist upon being heard, to maintain his independence. This obligation rests heavily upon engineers.

To be a professional engineer, in the true sense, does not require that we have some special set of relationships to society and to the organizations of which it is made up. It does require that the primary motivation be the acquisition of scholarship and its generous application to the needs of man.

To be an engineer in these days is to bear a proud title. To be able and willing to speak true opinions on the complex technical affairs of the day, without prejudice and free from control, is a privilege that is becoming rare in the world. Insistent upon his prerogatives, kowtowing to no man, respected because he speaks a truth the country needs to know, the independent engineer stands as an important member of the professional class—a strong bulwark against disaster, which can guide our steps into the ways of pleasantness and into the paths of peace.

10

THE QUALITIES OF A PROFESSION (1939)

Bush advances in this essay the creative concept of the engineer as minister, or "shaman," invoking the metaphor as a way of self-reflectively seeking to balance the centrality of technical authority with the obligation to serve one's community. Speaking in January 1939, on the eve of World War II, when Americans are trying to decide whether to mobilize for a global war, he presents a distinctive (and still relevant) way of thinking about what kind of profession is electrical engineering, and the nature of the obligations that electrical engineers owe society. The son of a respected Protestant minister in his native Massachusetts, Bush surmised that the engineer—and other emerging professions in an egalitarian country, such as scientist, doctor, lawyer, and teacher—had their roots in spirituality and the ancient search for meaning in a confusing world. "We can start far back" before recorded history, Bush began. "In every primitive tribe there was some sort of medicine man. He was a man apart, the adviser of the clan rather than its titular leader. He spoke, in his field, with authority, and this rested upon a special knowledge which he was supposed to possess. The medicine man was the progenitor of the professional man of today" and, in particular, "The descent of the engineer from the medicine man has been highly involved." Bush went on to trace "a central thread" running from the shaman to early priests in organized religions to the contemporary engineer. Spiritual leaders, like engineers today, lived by "a strict code of conduct." They "trained neophytes, subjected them to a long period of apprenticeship, initiated them into the mysteries, and inculcated in them pride in the cult, and rigid discipline in its formulas." As engineers do now, initiation into the field required surmounting "intellectual hurdles" and learning "a special language." Moreover, then as now, the shaman "sat as adviser in councils of the mighty. But more essential than all of these, he ministered to the people."

The concept of ministry is central to responsible, ethical engineering, Bush argues. "Ministry carries with it the ideas of dignity and authority," he explained. "It connotes no weakness and offers no apology. The word has been carried into diplomatic usage; and in the derived form of administer *into law and business. There is no fog of subservience surrounding the concept. The physician who ministers to his client takes charge by right of superior specialized knowledge of a highly personal aspect of the affairs of the individual. The attorney assumes professional responsibility for guiding the legal acts of his client and speaks with the whole authority of the statutes as a background. It is in this higher sense that we trace the thread of ministry to the people." And no less should be demanded of the engineer at large. "The great mission of the engineer," Bush insists, "lies in intelligent, aggressive, devoted ministration to the people."*

The ideal of "ministering to the people" isn't the animating principle behind required courses in university electrical engineering programs nor does it translate easily into a code of conduct for individual engineers. Bush's model for the social role of the engineer is revelatory and even revolutionary. His model is aspirational, situation-dependent, protean—and well-suited for times in which the stuff humans build destabilize traditional ways of life and undermine longstanding values.

The revolutionary road of course coexists and coevolves with what we might call, following the historian of science Thomas Kuhn, "normal" engineering. "Normal" engineers maximize stability and honor traditions through respect for "path dependence," a term which refers to those forces that keep things the same, or minimize change. Normal engineering emphasizes individual conduct and celebrates personal choices that are fair and responsible and reflect the integrity of effective technical practices. Where a particular engineer fits on the continuum between revelatory and normal requires study and depends on specific circumstances. Seeing around the bend in the river—or even knowing the river ahead does indeed bend—can require the sort of leap of faith that led Bush to believe that shamans, priests, and spiritual seers of long ago are the antecedents of today's celebrities and stars of techno-science.

THE QUALITIES OF A PROFESSION

Here I wish to trace briefly the relationship of engineering to the other professions, the professional traditions which engineers inherit, and the outlook for the engineering profession in view of its unique relationship to society. I plan to review the history of professions very sketchily; but through this history runs a thread to which I wish especially to direct attention.

We can start far back, but not tarry long in our review. In every primitive tribe there was some sort of medicine man. He was a man apart, the adviser of the clan rather than its titular leader. He spoke, in his field, with authority, and this rested upon a special knowledge which he was supposed to possess. The medicine man was the progenitor of the professional man of today.

His closest modern counterpart is the scientist. The scientist and the medicine man have much in common. Tribal regalia and feathers have undergone metamorphosis and reappear in cabalistic titles and letters surrounding names. The queer jargon of the cult has changed in nature but preserved its hypnotic effect. Solemn pronouncements about the unknowable still catch the ear of the multitude. The claim to a favored position in society is still based on the occasional ability to unscrew the inscrutable. In fact, one difficulty that faces the scientist is that he may be mistaken for a medicine man, able at will to produce rabbits from hats, instead of the careful, hard-working human individual he really is.

The descent of the engineer from the medicine man has been highly involved; and it will clarify some obscure relationships if we trace part of it, for there is a central thread which runs through the tale.

The medicine man, and the member of the pagan priesthood which succeeded him, was characterized by numerous attributes. He had a strict

The text is from an address by Bush before the American Engineering Council, January 13, 1939. Reprinted with minor changes in Vannevar Bush, *Endless Horizons* (Washington, D.C.: Public Affairs Press, 1946).

code of conduct. He trained neophytes, subjected them to a long period of apprenticeship, initiated them into the mysteries, and inculcated in them pride in the cult, and rigid discipline in its formulas. He severely restricted his numbers, by intellectual hurdles to be surmounted. He spoke a special language. He sat as adviser in councils of the mighty. But more essential than all of these, he ministered to the people.

This was the first professional group and all others have derived from it. Not every attribute has been maintained as new professions have emerged, but to a surprising extent their counterparts can still be found. In every one of the professional groups, however, will be found the initial central theme intact—they minister to the people. Otherwise they no longer endure as professional groups.

Ministry needs definition for our purposes. The alteration of word meanings with new usages is such that it is only too easy to misunderstand. Ministry is not service, and we have so completely altered the essential significance of the latter word that it may have utterly different connotations to different hearers. Ministry carries with it the ideas of dignity and authority; it connotes no weakness and offers no apology. The word has been carried into diplomatic usage; and in the derived form of *administer* into law and business. There is no fog of subservience surrounding the concept. The physician who ministers to his client takes charge by right of superior specialized knowledge of a highly personal aspect of the affairs of the individual. The attorney assumes professional responsibility for guiding the legal acts of his client and speaks with the whole authority of the statutes as a background. It is in this higher sense that we trace the thread of ministry to the people.

This is the fuel which has kept alight through many ages the professional spirit. Every time that the fuel has become exhausted, the light has gone out. It has not mattered how much was retained of the trappings and mysticism, nor what the profundity of utterances, there has been no true profession that has not with dignity and authority advised and counseled the people that has not guarded the commonweal. For a true profession exists only as the people allow it to maintain its prerogatives by reason of confidence in its integrity and belief in its general beneficence.

The monastic orders, under diverse religions, springing up as outgrowths of the simpler systems of the priest craft, have exemplified the theme in two ways. Some have preserved, adorned and extended the knowledge of their time and place. These have their modern counterparts in the scientific and learned groups, the custodians of our culture, and the source from which flows new knowledge for the use of man. Other orders carried to great heights the direct ministry to those in misfortune or distress—often great self-sacrifice as did the early Jesuits among the Indians of our West. Both forms have remained high in the esteem of the people and have endured. Occasional groups have lost the thread, and have, for example, become militant orders devoted to self-aggrandizement, and these have disappeared.

Out of the early priesthood came also the teaching orders, whose ministry took the form of instruction of the young, and this aspect of professional activity is represented today by the great profession of teachers everywhere. This group has little indeed of the trappings of the medicine man, it has no single closely knit aggressive society representing it; the language is becoming complicated but is still fairly intelligible to the layman. Where it has maintained its ideal it is honored and respected. Great teachers do not find riches heaped upon them. They do not become affluent. Great teachers have no interest in riches. In the great teacher the paternal instinct, which is so often at the basis of senseless extremes of individual striving for wealth, becomes sublimated into a broad love of youth which calls for neither wealth nor power for its enjoyment or satisfaction.

A very early offshoot was the profession of medicine, for ministry to the ill was a primitive need. It has had a long and distinguished history. Utilizing the fruits of science, it is today in full tide of accomplishment for the benefit of mankind. It has had vast power and influence. Yet today in the United States [medicine] is at a turn in the road, and its most thoughtful members are giving earnest consideration to its future. There is serious danger that its light may fail, and its heritage of idealism may be lost.

The profession of medicine, by reason of its very nature, has preserved many of the attributes of the ancient forms. It selects its neophytes by rigorous intellectual elimination, trains them over many years and seeks to endow them with the philosophy of their profession. It severely restricts its own numbers, perhaps too severely in view of the task before it. It preserves itself apart, by special language, and has a unique code of conduct. It has sat in the councils of government and advised. By the will of the people, it has given special privileges and prerogatives for use in pursuit of its objectives. [Medicine] is highly organized. Through long ages it has held well to its ideal of simple ministry to the people and has disciplined under its codes those who would use its special privileges for other ends. It has guarded the people against their own folly and has been properly militant in maintenance of its sphere of the public weal. Its individual members are in general respected in their own communities to an extraordinary degree.

Yet in these days when all institutions are undergoing scrutiny, when our population in fear and distress is prone to be critical, there is evidence about us that the profession, as a profession, does not command that full support of the people of the country without which it cannot continue on the path. Yet as one looks about in the medical profession, signs are seen of a resurgence of idealism, a re-emphasis on the simple mission of healing and a recognition of the central theme of ministry to the people. This I am convinced is the true motive of the great majority of the members of this grand profession. Yet there is much suspicion in the public mind that aggrandizement, utilization of power for the professional advancement of the membership, the guild spirit in its cruder form, are rampant. I second the thought of many eminent members of the profession itself that unless this suspicion is allayed by a revival of simple ideals, the profession will suffer and the people will suffer enormously with it. It is well that engineers should be deeply interested in the outcome for medicine is a very ancient profession from which we have much to learn.

To treat the origins of the profession of law, its codes and countercurrents, would require an article in itself. Here is a field in which the preservation of the true philosophy of a profession is intricate indeed. Endowed with special privilege under the law, it largely regulates its own conduct. Never quite successful in the recruiting, training and indoctrination of its neophytes, its maintenance of adherence to a high code of conduct is rendered more difficult. Counseling with government and, by nature of its mission, participating directly therein, it has great power for good or evil. It certainly strives, as an organized profession, for the public welfare. But its zeal in this regard is not always such as to cause it to disregard the special welfare of its own group. And the two are sometimes hard to disentangle. [The legal profession] ministers to those in legal distress with great effectiveness but the distressed often appear in pairs. It is hardly judged as a whole by the public. Certain it is, however, that those of its membership, on the bench or at the bar, who have risen to the highest positions in their devotion to professional ideals, are respected and honored by the public. Certain it is also that, should this respect falter, we as a democracy would soon be in a sorry state. But our principal concern here is the engineering profession and, we inquire, what is the engineering profession? Is it a profession at all? And if [engineering] is, will it develop into the full stature to which the importance of its work entitles it to aspire?

[Engineering] is relatively young. The military engineer appeared in the first steps of the mechanization of warfare, when forts began to take shape. His counterpart in peaceful affairs was called a civil engineer. With the industrial revolution, and especially with the spread of mechanization from the factory into every walk of life, engineering became exceedingly diversified. Applying science in an economic manner to the needs of mankind is its broad field. Its disciplines are spread over all the sciences as they become thus applied, and embrace also portions of economics, law, and business practice, which are integral parts of the process of application. It is somewhat loosely organized as professions go. To a minor extent only, [engineering] limits its numbers. But the very strictness of its essential disciplines provides some selection of its neophytes. Until recently, it has done very little in an organized fashion to inculcate in its younger members the philosophy of the profession, leaving this largely to those of its individuals who are also members of the teaching profession. That branch which represents the consultant, and others to a degree, has drawn codes, but there is no body of codified principles which is accepted and applied by the profession as a whole. [Engineering] has no highly distinct language or jargon for it must continuously work with laymen.

These are, however, incidentals. The important point is this: does [engineering] have a central theme of ministering to the people? Most certainly it serves the public in myriad ways, but are its individual members activated primarily by the professional spirit of dignified and authoritative counsel and guidance?

In order to properly inquire into this weighty question, we need to digress a moment to consider another large group of the population, the modern men of business who have derived from the ancient traders and merchants. The merchant class has not usually been a professional grouping in the true sense; and engineering, which has derived its philosophy from this source as well as science, naturally partakes of the heritage of both groups. Business has served the public of course, but its main theme has been the profit motive, a salutary objective when restricted by law to the use of ethical procedures in its pursuit, but not a professional objective.

One of the most encouraging signs of the times is the gradual emergence in our day of the truly professional man of business. Scattered, not organized, with no sign of professional trappings, they are nonetheless possessed of a high mission, which needs only formulation and recognition in order that they may constitute a new and strong profession. This is occurring for one reason—because of a gradual change in corporate form. The owner-manager was activated by the profit motive, and no amount of paternalism could wholly alter his position in the social scheme. Even with corporations, the ownership of which is highly scattered, the manager is ordinarily controlled by and primarily responsible to a few powerful owners, so that he in essence still represents the interests of the owners in his relations with the three bodies with which he deals: the government, the employees and the consuming public. It is his difficult task to reap for the owners' benefit the fruit of his industrial operation, while maintaining at least tolerance on the part of the other bodies. But there are some corporations in which ownership is so diffused that the management becomes in effect a self-perpetuating entity, partaking therefore of the nature of a trusteeship, with equally weighed obligations and responsibilities to all four bodies: owners, employees, government, and consumers. Among such managing groups will be found individuals who have the professional philosophy in high degree, conducting their affairs for the just and equitable benefit of all four groups concerned, maintaining the health and progress of their institutions as potent agencies for ministration to the needs of the people. They find common ground with the trustees of great foundations, of hospitals and all non-profit organizations devoted to the public welfare. They find common ground also with many of those who make a career of the business of government. Their ranks are recruited by many in the ordinary walks of business who have seen a light and envision a function in life which is higher in its satisfactions than the struggle of anybody against any other; namely a struggle with all bodies to preserve an ideal. Out of this trend, as competition for industrial existence becomes tempered, should emerge a new profession with its own traditions and beliefs, which is capable of managing prosperity so that it will be conducive to the health of a nation. And there is a grave question whether this objective can be attained in any other way. I wish there were a special order of knighthood in this country to honor and unite those who are now blazing the difficult path and developing the novel philosophy of this new profession.

Engineering, however, derived jointly from the quiet cloisters of science and from the turmoil and strife of aggressive business, and it is no wonder, therefore, that it should wobble a bit as it seeks to evolve its own professional philosophy. Just as it is not reasonable to expect the young neophyte to grasp at once the idealism of his calling, so it is perhaps not reasonable to expect a profession which is so young and which has grown so fast to have found itself in this regard.

The period of initiation into any profession should extend into maturity. Only when members reached the full bloom of manhood did the ancient orders entrust the mysteries to their care. The young neophyte served his apprenticeship under constant tutelage and close guidance by mature mind, and this we still find in every profession. As apprentice, as employee, he is called upon to prove himself before he enters into that relationship where his opinions are controlling in his special field, and some there are who never emerge from close control and the mere exercise of technical proficiency. In the engineering profession this emergence usually is circumscribed by the fact that most engineers operate as members of industrial organizations of one sort or another, and the fact that they serve their apprenticeship in this same sort of organization and come to devote their entire efforts to its affairs, rather than to enter them after professional recognition elsewhere, as is usually the case with medical or legal individuals. This, however, merely emphasizes the need for better supervision of the neophytes by the members of the engineering profession who have arrived. It is not enough to leave their training to the industrial organizations of which they are junior members. Inculcation of the principles of the profession can only come from those who themselves have attained to a full grasp of its proper function in society, who have arrived at a balanced judgement as to its responsibility to the several groupings of which society is composed, and who have a professional interest in the young men who are destined to succeed them in the profession. Every profession should have its secrets and its mysteries spread before the world that all may read, but truly grasped only by those who have lived the professional life. And these [secrets and mysteries] should be transmitted to the neophytes with due care, with reverence for their inherent worth, and in due time. Ritual and symbolism, secrecy and circumspection, were the ancient paraphernalia which insured a proper seriousness in youth in order that the impartation might be impressive. These have not wholly disappeared from modern professions. Admission to the bar, the use of the title of doctor, and similar customs and usages have profound effect in producing a professional consciousness. The engineering profession is wholly without these aids, and its task of inducting its neophytes into the true professional atmosphere is thus rendered doubly difficult.

But does it matter after all? Are the things that engineers do so vital that they must needs be approached in the professional spirit? Most certainly it matters. And most certainly the task is a professional one. The impact of science is making a new world, and the engineer is in the forefront of the remaking. He lights the way in a very literal sense. He brings people close together for better or worse, by facile communication and rapid transportation. He guards the food supply and replaces the hopelessness of Malthus with an embarrassing plenty. He shortens the hours of labor and fills the consequent leisure with distractions. He temporarily disrupts the techniques of whole industries and thus alters the life habits of many people—[while] maintaining a continually rising standard of living. He bores through the earth and under the sea, and flies above the clouds. He builds great cities and builds also the means whereby they may be destroyed. Certainly there was never a profession that more truly needed the professional spirit, if the welfare of man is to be preserved.

There is no lack of signs of a rising consciousness in this regard. The profession is most positively vocal. There is a vigorous new organization, linking several large groups, devoting itself to the improvement of the education of the young engineer, and the instillation of high principles during his early career. Engineering literature is full of discussions of the duties and responsibilities of the profession, and out of this may crystallize someday a code, a set of principles of conduct, a guide drawn solely with the object of advancing the public weal, which will become accepted by engineers everywhere, whether in government employ, private practice, or industrial organization. Having, to some extent at least, consolidated their techniques, engineers are certainly giving thought and voice to their position in society, and to their responsibility for the use of the great works which they create.

The focus of this whole affair is the American Engineering Council. More than any other group, it represents the engineering profession as a whole, in its relationships with government, other professions and the public. Here, more than in any other organization, reside the external as contrasted with the internal relationships of the profession. It was founded by men who considered its functions in terms of a high idealism. It is now going through a strenuous period of self-examination. To this every individual can contribute only one set of thoughts to be merged with all of those which seethe, and interact, out of which will come in due time that consensus which will form the opinions, traditions, codes and consciousness which will mold the engineering profession. It will come unless the Council fails; for if it fails, and if its place is not taken by a more rugged successor, there will be no unitary engineering profession at all. In the spirit of adding my few thoughts to those of the eminent men who are directing the Council, I have previously offered comments and now I comment again, with the expectation that I will be disagreed with and answered, with the wish to add my [modest contribution] to the consummation.

I find it a vigorous and rapidly evolving body. I consider it to be utterly inadequately supported by the profession as a whole, in comparison with the central bodies of sister professions, and with a serious problem as to how adequate support can be drawn for the great task that lies ahead of it. I find it partaking of the great American tendency toward over-complication and inclined to attempt things which seem to me personally to be off the main beat. I find to my great joy that it is gradually becoming known and recognized; and I trust this is just a beginning. I find it guided by some of the best minds in the profession as its officer, who are giving valuable time to its cause; and I hence cannot fail to be optimistic as to its future.

To me, however, there is just one point on which I wish to focus attention. I find it struggling with its own philosophy. I find, in fact, that it hesitates as it formulates its idealism; that it has not yet placed its foot unequivocally and irrevocably upon the path that leads to complete devotion to the public welfare. I find that it has not yet enunciated its belief that the great mission of the engineer lies in intelligent, aggressive, devoted ministration to the people. This I will urge with all the emphasis I can command.

I do not seek to conjure away practical difficulties by ignoring them. I know full well what restricted budgets mean. The argument that the support of the membership can be obtained only if they can see a direct and personal benefit from their contributions has a familiar ring. I recognize that it is entirely proper for professional men to join in an insistence upon a reasonable and proper recognition of their services to society. Yet if there is no central organization which has as its creed the best service of the profession to the society of which it forms a part, then there will be in the end no engineering profession. Professional status rests in perpetuity, not on transient law, not on the cruder mechanisms of the ancient guilds, not on exclusive control of those having a specialized and necessary knowledge; but upon the respect and fundamental support of the people who are served, who only in the long run can insist upon the maintenance of prerogatives, and confer honor, recognition, and special privileges in society upon the members of a profession.

Will engineers support such a program? Will they contribute directly or through their specialized societies to the development of this ideal, and its exemplification in Council projects aimed at rendering real some aspect of the profession's contribution to public welfare? Will they make possible great forums for the crystallization of engineering opinion on public questions involving engineering, not to attain an impossible unanimity or produce high-sounding resolutions, but so that all aspects of controversial matters may be aired in order that people may know what engineers think? Will engineers go along heartily in developing a professional consciousness, a code of action, a philosophy which implements a desire to be a truly professional group, oriented primarily toward the advancement of the public health, safety, comfort and progress? Will they accept as the central theme the engineers' ministration to society, without fear of class, and without prejudice toward or away from any special social interests or causes?

If they will not, then there is no truly professional spirit to build upon. We may as well resign ourselves to a gradual absorption as controlled employees, and to the disappearance of our independence. We may as well conclude that we are merely one more group of the population, trained with a special skill, maintaining our economic status by a continuing struggle against the interests of other groups, forced in this direction and that by the conflict between the great forces of a civilized community, with no higher ideals than to serve as directed, and with no greater satisfaction than the securing of an adequate income as one member in the struggle for the profits of an industrial age.

But I know the minds of too many engineers to be thus pessimistic. I recognize the distinguished careers of a generation of men who have practiced in the profession to its credit and honor. Though the task be difficult, on account of the nature of many of our relationships to society, nevertheless traditions are being formed, the consciousness of the membership is becoming aroused, and I confidently expect the profession of engineering to develop in a manner of which we can be justifiably proud.

And to those in the ranks who may not have yet seen the light, I would preach the doctrine, without pulling any punches, without mincing any words, that the path of professional attainment lies open before them, that it is a thorny path that is easily lost sight of among the rocks and rubbish, that it can be adhered to only by sacrifice and by support of those who lead the way, that it is a long path which leads down into the valley into which the sun does not shine, but that it leads at last to the heights—to the heights of true professional attainment, where honor and individual recognition by fellows is the real reward, and where the watchword is that old, old theme, which has never lost its power, and which may yet save a sorry world, simple ministration to the people.

11

INNOVATION, ENTERPRISE, AND THE CONCENTRATION OF ECONOMIC POWER (1939)

Economics, business, and innovation were among Bush's enduring interests, and his insights into these subjects ranked among the finest provided by any American in the first half of the twentieth century. The Depression created profound confusion among professional economists of the era, and the persistence of the Depression raised grave doubts about the capacity of government to sustain economic growth and employment. Bush was among a relative few independent thinkers who saw in technological change the source of rising future prosperity for the nation and rising living standards for his fellow Americans.

Drawing on his own experience founding an electronics company, Raytheon, in 1922, Bush saw the potential for new processes and products, sometimes birthed by small firms or lone innovators, to expand jobs and consumption. His enthusiasm for commercializing innovations—and the role to be played in this process by what he called "the independent man"—forced him to learn about the ways in which the American legal system protected intellectual property through a system of patents and copyrights. As the Depression deepened and persisted in the 1930s, some observers began to believe that economic concentration, largely in the form of corporations holding monopoly power over important markets, was strangling America's innovation potential. Large companies were using their patents and their legal resources to

*protect themselves from new competition brought about by technolog-
ical advance. Bush himself had seen this problem up close when RCA,
which dominated the design and manufacture of radio "tubes," challenged Raytheon, which made competing radio tubes, in court. The legal
battle proved costly, though Raytheon survived. Because of his experience in business, and his stature as a university leader and a designer of
what were then among the world's most powerful (electro-mechanical)
computers, Bush was asked to testify to a joint committee of Congress
that held nearly three years of public hearings on the costs of economic
concentration in the United States.*

*The hearings were held against the backdrop of arguments over
the causes of high unemployment and economic contraction, and the
nagging belief that the Depression persisted because of either too much
technological change (notably in the form of labor-saving machinery
that seemed to reduce overall demand for workers) or too little change
(because large economic actors had no incentive to introduce potentially disruptive advances into the market). Into this debate strode
Bush who, while not an economist, spoke plainly and confidently about
these subjects. In this selection, Bush addresses various subjects central to arguments of how engineering and science, research and discovery, interact with the engines of capitalism, finance, and markets.
He is especially keen to explain the importance of new enterprises and
how small companies can grow into large ones on the strength of new
ideas, applications, and markets. Many of his ideas anticipate later
conventional wisdom about how government support for research and
discovery sustains economic prosperity. Bush's remarks, which came as
part of his lengthy testimony to a joint House-Senate committee concerned with what Bush succinctly describes as "an appalling problem of
unemployment," represents the clearest public expression of his ideas on
innovation, the law, politics, and economic growth. The skillful questioning of his interlocutor, John Dienner, an expert on public policies
and laws around the economics of research and the commercialization
of ideas, helps bring out the style and substance of Bush's visionary perspectives. What follows are paraphrases of Dienner's questions, followed
by excerpts from Bush's testimony.*

INNOVATION, ENTERPRISE, AND THE
CONCENTRATION OF ECONOMIC POWER (1939)

What's the value of the "pioneering spirit" to American technological advance?
There is not the slightest question that this country has a high standard
of living as compared with other countries. That has been brought about
for several reasons. First, this is a country of pioneers. The frontiers have
disappeared geographically as the frontiers of technology have advanced.
Pioneering experience still remains to a certain extent. The pioneering
spirit, that willingness to take a chance, has been very important in our
industrial advance.

How do new ideas come forward and what produces them?
There are two ways that are important. First [ideas] result oftentimes
from the long program of research, careful and meticulous analysis of
the situation by a group of men, through industrial research laboratories
or scientific institutions and the like, which produce new knowledge out
of which come new applications. In addition, there is the independent
inventor, whose day is not past by any means, and who has a much wider
scope of ideas and who often does produce out of thin air a striking new
device or combination which is useful and which might be lost if not for
his keenness.

What is research?
Research is broadly the discovery of new knowledge by systematic examination and it can be classified on one basis, into pure research, applied
research and research for control of a product.

*Who carries basic and fundamental scientific research and how does funda-
mental research differ from applied research?*
That is carried primarily in our great institutions of learning, in our academic institutions, universities and the like, and also to a very considerable

Vannevar Bush testimony before the Temporary National Economic Committee, also known as the
"Monopoly Committee," January 17, 1939; excerpted from pages 869–911 of the Hearings Before the
TNEC, 75th Congress, third session.

extent in industry itself, for the great research laboratories in this country in industry carry on basic research as part of their activity. . . . There is considerable difference. . . . In pure research, basic research, men are left comparatively free to follow their own ideas. In applied research, they are of necessity guided in the direction of interest of the company which employs them.

Are there limitations on new ideas and further research?
In the first place there is no limit to the new ideas that can be produced. We are not at the end of industrial advance, we are not at the end of scientific advance by any means. New ideas are coming forward with as great a frequency today as they ever have, and while a great research laboratory is a very important factor in this country in advancing science and producing new industrial combinations, it cannot by any means fulfill the entire need. The independent, the small group, the individual who grasps the situation, by reason of his detachment, is oftentimes an exceedingly important factor in bringing to a head things that [would] otherwise not appear for a long time.

Does any role remain for the individual in the innovation game?
I too have wondered whether . . . the individual is disappearing. Personally, I don't think he is. Certainly in pure science he is not. In pure science today the individual can map his own path and make his own recognition as an individual . . . As I have tried to bring out, [there] is one phase of the production of new ideas, a very new and I think [beneficial] phase, a group phase, but the individual phase has not disappeared and there are still in this country plenty of individuals with ideas which are important which ought to go into industry for the benefit of the people of the country, produced not by group work but simply by reason of the fact that there are individuals who have that keenness of analysis, of grasp, which enables them to see long before anyone else in the population a trend and a need, and to put together a combination or a device which will satisfy it, and we need those people. They have been very important in the advance in the past, and we need to facilitate their action in the future.

Why unemployment despite industrial expansion?
While industrial development in the U.S. has been perfectly tremendous and marvelous, nevertheless it is accompanied by an appalling problem of

unemployment, which again indicates that operating on the new frontiers we have not been able to do what the pioneers in the days of geographical frontiers could do; namely, find a way of supporting the individual properly.

What is the value of new firms, and the virtues of small organizations for innovation?
We want to be careful that we do not confuse research laboratories with *large* research laboratories. Now I remember the research laboratory with which I was associated in the early days which produced, it happened, a new industry. It consisted of four of us and we were a corporation; but it was essentially an individual effort for bringing into use some ideas. Now every industrial affair was once small, and I think my own attention is particularly on the point [about] . . . the need for facilitating the progress of these small things which may grow into large ones.

What's the role of universities and other non-profit organizations in maintaining quality research and research laboratories?
The great academic institutions of this country of necessity maintain a great deal of scientific and technical research, for the simple reason that the highest form of instruction, the highest form of teaching in its advanced stages, can be given only in the presence of research. They necessarily extend the frontiers of knowledge at the same time they are teaching, so that you will find all of the better academic institutions of this country doing research within their corridors. And then there are in addition organizations that are non-profit . . . endowed [and] formed for the simple purpose of advancing knowledge, such as the Carnegie Institution of Washington [which Bush heads].

How do other nations conduct research and how does the pace compare with the United States?
This is a world of competition. I think that if we are to hold our position in a competitive world, we need to be in the forefront of science, we need to be in the forefront of applications, and we can do so only by having the facilities for research and more important of course the people for research, the young people who are trained and able to work in that field.

What's the competitive challenge from Russia?
I was in Russia about 11 years ago and there every piece of research, every laboratory, every individual working in science or in its applications, is very definitely controlled by the needs of the central government and their interpretation to him. He is directed and definitely in the lines in which it is desired that he should function. . . . Of course the Russians have produced great scientific things and in recent years. In mathematics, for example, they have done excellent things, and some of their men in pure science are given the freedom to enable them to do such things, but the industrial research and the research generally is closely directed . . . and under very definite control. We do not have the independent man there [in Russia] as we have in this country.

Are America's research organizations available for emergency needs, such as a war or an epidemic?
In regard to the first, if we get into another major difficulty, one of the primary things that we would need is a group of trained and able individuals capable of advancing the means of warfare, and I regard it as highly essential, as a part of our national defense, that there be encouraged in this country research laboratories of all kinds, the training of research personnel to a high degree in order that they may be readily available if they are needed in an emergency. . . . I think it applies to every field. Biological, chemical, electrical fields. Of course in the matter of public health a great deal may be said. We have come far in this country in medical research, and the progress has not stopped by any means in this regard. We are not beyond the time of possible epidemics. We may again meet difficult problems in this country in epidemics, and if we do our resistance to those will depend upon the skill and number of organizations and men in medical research and allied practices.

Can rival companies in the same industry "pool," or share, patents in such a way as to suppress competition and reduce innovation?
The idea of suppressed patents may take several forms. One form that is fairly frequent is this: A company has two ways of accomplishing the same thing. It has two patents, either one of which it might use in producing a device for a given purpose, and it may produce one of those and not the

other. I do not personally regard that that is a suppressed patent provided the public need is meet.

Another form in which I have heard the term: The advent of inventions, the advent of industrial devices, is sometimes delayed because the company which controls the patent situation thereon does not produce the devices for the public use as rapidly as it might. That is again a matter which can't be settled in a moment, can't be dismissed in a word. Sometimes it is economically desirable that the obsolescence of equipment in the hands of the public be brought about deliberately and reasonably gradually, and not abruptly and suddenly, for sudden obsolescence would produce disruption, unemployment and what not, so that I think oftentimes delays of that sort are justified. . . .

There is no great danger of leaving that judgement [on when to market a patented innovation] in the hands of the company itself, for this reason: That this is a temporary monopoly which the company holds, and if it delays unduly it destroys its own monopoly because the patent is going to expire. . . .

I have never seen a suppressed patent [for the purposes of protecting an inferior product or service]. And I think the reason is this: It is altogether too dangerous a procedure.

Let's take the example of the vacuum tank on the automobile. The patents on that system were pretty much held by one company which controlled the system of transferring gasoline from the tank to the engine by a vacuum device, and they had a group of patents that controlled that whole affair.

Whether or not they put into the hands of the public the best form of that I can't say, but certainly they had an incentive to put out the best form of it. Moreover, they were vulnerable. One would have said offhand that they had the entire situation in their hands, but what occurred? Along came the motor gasoline pump and the vacuum tank became obsolete.

Every company is in the position, even if it has control of a particular device, that some individual may come on with a new and novel idea which will render their whole affair obsolete. The more complex their situation in some ways, the greater the danger, and they have therefore the greatest incentive to make the best device that they possibly can in view of the things that are in their hands, and my own experience and my own judgement is that there are no suppressed patents in that sense and that it would be very foolish for industrial concerns to have suppressed patents in that sense.

Does a patent, though temporary, effectively confer on its owner a permanent monopoly?
I will take a moment to discuss that because it can't be answered in one sentence. The original patent law contemplated an original inventor and gave him a monopoly for 17 years, after which the monopoly would terminate. That still happens today. We do have individual patents which stand on their own feet and which are used for 17 years by the original inventor or his assignees, and then go into public use. But we also have various other situations, and one which is fairly clear-cut is this: The company has a group of inventions protected by patents. It has intensive research and as a result it continually improves its product and takes out new patents and in that way extends its monopoly.

My own point is this with regard to that particular form: if a company can improve its product at the rate necessary to preserve its patent control, assuming again a reasonable expediency in prosecution and that we have no long delays, then I say that is for the public benefit. It is a monopoly made permanent for a time by reason of the activity of the holder thereof. It is bound to expire sometime and it is in general beneficial because of the incentive which that company has to greatly improve its product. . . .

And that brings me to the third form. It is not easy to prevent some individual from making the improvements, but it may readily be that if that individual makes such an improvement, he will find himself in the position of having only one customer for it, so that an organization which has a patent control over the entire situation may therefore find it readily possible to acquire improvements which come from the hands of others and to thus perpetuate, by aggregating to itself the improvements not only that it itself makes, but also the improvements made by others by purchase. . . .

Our entire patent system is based on the idea that any individual, if he thinks he is right, may make anything, and the recourse of the holder of the patent, no matter who he may be, individual or corporation, is to appeal to the courts . . . it would be a very foolish thing to do, for a company which has an established business will certainly not go out and boldly violate a patent which is obviously valid, for it is so vulnerable that that becomes exceedingly dangerous. The individual, through the courts, can collect damages. . . . No company would go out and baldly violate an obvious patent in the hands of an individual.

Are specific patent pools undesirable or objectionable?
The simplest situation that arises is this: where two companies or two individuals hold the patents and neither one of whom is able to manufacture on the basis of the patents which he holds, so that it is necessary for them to get together in some way or other before the device can go into public use. Obviously in such cases it is to the public interest that they should interchange rights under the patents. That is the simplest situation.

We should have a more complex situation, however, where units in an entire industry interchange patents, and we have then what we might call a patent pool. In my opinion some kinds of patent pools are necessary and beneficent, and other types are undesirable. . . . One undesirable feature, I think, is this. If the patents are interchanged among the units of an industry on the basis of no royalty, I think that is undesirable, because the incentive which is provided by the patent system for progress, for research, for invention is effectively canceled out in that event.

I think also that a closed pool which has no provision for the entrance into it of a newcomer, who brings with him an addition to the situation, is undesirable. I wish very much that a beneficent type of pool, a desirable type of pool, could be defined and given public support or governmental support in this country, for I think it is a thing that we very much need. Pooling is necessary and desirable if properly carried out. . . . [To insure they are properly carried out] reasonable royalties should be charged between the units of a pool in order that the incentive to progress be not cancelled out. . . . I think in general that pools are very necessary in some fields . . . and should be encouraged. Of course, the converse is also true: that I believe that pools very often have been disadvantageous in the past where they have not contained the desirable features.

Is there value in allowing the U.S. government to compel companies to license their patents for modest royalties in order for others to more quickly

create beneficial products and services? Will corporations and private inves-
tors abandon research if its fruits must be shared widely with others?
Some of them would [halt research]; some would not. Some great research
laboratories have other purposes than the mere production of patentable
inventions. The research laboratory of the telephone company [AT&T],
for example, has many other functions and many other ideas. I can tell
you that many would never come into existence and many research laboratories and many groups now and in the recent past striving to bring in
new products would never have come about had there been any system
of general compulsory licensing. I can tell you from my own experience
that I was closely associated with the founding of several small companies
in this country based on inventions, and no one of those would ever have
come about had there been a system of general compulsion in licensing,
so that having spent a great deal of money [developing their ideas], they
would have been obliged to license their competitors at a small royalty. . . .
Some things would come into use [without the profit incentive] but there
are many ideas for which it is essential.

How important is what Bush describes in his testimony as "venturesome
capital" (and what today would be called "risk capital" or "venture fund-
ing") to the health and success of technological innovators?
The introduction [to the market] of an invention requires a large initial
investment. Then funds for that can be secured only if there will be a
speculative profit, only if the individual who puts up the money can expect
that if the gamble is successful, he will reap considerable profits. . . . [Thus
is] the general situation in regard to the attraction of new and venturesome capital into new ventures.

Can government effectively assist technical innovators, who Bush called
"pioneers"?
If we can remove some of the difficulties in the way of the pioneer, the
technical pioneer, if we can make it more readily possible to establish
new industries in this country based on inventions, if we can remove
some of the difficulties of litigation [over patents], if we can simplify the
procedure [of obtaining "intellectual property" rights], then I think we
have a reasonable chance that we can regain our [leadership] position

and proceed on the way. Unless we do that our industrial progress will be permanently lost.

Is technological change and scientific advance slowing or growing?
I expect an acceleration. . . . Science has gone ahead at an accelerated rate. . . . The progress of the world is not stopped in any degree. . . . The extent of human desires in infinite. The extent of human needs may be bounded, but there is no limit to the number of new devices and new advances that can be absorbed by the public if they are produced at a reasonable cost and properly distributed, and we are nowhere near the end of new devices for the public benefit, new combinations, so that I fully expect the program will take two forms: the production of more of the usual things that we already have and in better form by better methods and the introduction of wholly new things.

Invoking the concept of "creative destruction," without explicitly referring to the Austrian economist Joseph Schumpeter, who coined the term, Bush answers the question about the downside, or negatives, of technological innovation by conceding under questioning that "new ways of doing old things," can cause "at least temporary disturbance."
Progress, sir, always pays for itself by at least temporary disturbance. If we have a static world, we can have a completely stable affair in which things do not change. That is very lovely in one way, but if we are going to go ahead technically or in any other way then we must expect at least local disruption and temporary disruption, which means unemployment. There is no question, however, that the whole trend of invention, the whole trend of the introduction into industry of new devices and new ways of doing old things has been to greatly increase employment in the long run, and in the end, so that it works both ways. [Innovation] produces in my opinion a temporary and local disruption but in the long run and over a considerable period generally increases the standard of living and increases enormously the potential employment . . . As industries grow old, there must be the advent of new industries to pick up the slack or we will have difficulty. . . .

We have a race between the tendency of old industries always to produce their products with less labor and the advent of new industries which are capable of picking up that [displaced] labor and labor in addition.

What drives capitalism, wealth-creation, and rising living standards for many?
Only men of large means can properly take the long shot that is involved [in bringing breakthroughs to market].

The combination [of] the man of large means and the man with an idea [engender material advance]. I would tie it to both, for I think that only when you have the proper combination of the man with the good idea, the new thought, the new invention, and capital able and willing to enter into its development with it have you a combination which can produce new industries. . . .

Taking into account the whole situation as I see it, I think the courage and resourcefulness called for today in a man who would break new ground in the industrial field, produce new companies, new products for the benefit of the public and the risks that he takes, are as great as the risks of any pioneer; and his reward ought to be commensurate with the risk that he takes. . . . [In technological innovation] the day of the pioneer is not past, the day of the individual inventor is not past, for fine as these cooperative [research] groups may be and necessary as they are to our general progress in this country, they do not cover the entire field.

12

LETTER TO HERBERT HOOVER ON "THE WHOLE WORLD SITUATION" (1939)

Bush was not openly politically partisan, and in letters and his own reflections in the 1930s, he never felt the need or the occasion to express support for the Republican Party or denigrate the Democratic administration of Franklin D. Roosevelt, or the New Deal broadly. He clearly admired Herbert Hoover, Roosevelt's predecessor, and not only because Hoover's preference for self-reliance fit Bush's own sense of the value of individuality and independence from government rules. Like Bush, Hoover was trained as an engineer, and, also like Bush, Hoover conceived of politics as a method to mobilize citizens and society to address urgent problems. Further, Hoover served on the board of the Carnegie Institution of Washington and supported hiring Bush as president. Hoover lacked the willingness to address international tensions and the prospect of world war. And Bush's letter indicates his shifting perspectives on America's role in the world. Younger and more open to novel ideas, Bush had bold ambitions for engaging contemporary affairs through a muscular planning by the national government, as this letter reveals. Prefiguring his offer to President Roosevelt a year later to mobilize civilian scientists and engineers in the service of accelerating advances in new weapons and military equipment, Bush shares his ambitions with Hoover, saying that he wishes to personally help in the defense of the nation and, he suggests, he can "perhaps" succeed if he can tell "the right story to the right people."

For Bush, Hoover was a role model and inspiration—even long after he left the presidency. Four years later, in April 1943—at the height of World War II—Bush again wrote Hoover to emphasize that the former president's once fashionable isolationism no longer held up and that U.S. involvement in the affairs of other countries, especially those defeated in war, must be accepted, and for a very long time.

"I believe we have got to accept the policy of interfering with the internal affairs of the conquered [nations] for some time to come, on the same basis as interference with the affairs of backward peoples on a probationary status is inevitable," Bush explained in his 1943 letter, adding: "We have to have force in the world, the whole problem is how to have it responsive to the best will of all the people. I feel that this is the very core of the problem, and that it has not yet been completely thought through, even philosophically." What Americans must avoid in the future, Bush insisted, is to so abhor "the use of any sort of force" as "to allow the storm to gather and break in full force upon nations that are existing in a state of wishful dreaming." In short, Bush argues, "the severe realities of a world of ambitious men" demands an end to appeasement and the formation of "an effective police force." Perhaps "for the next thousand years," he concludes, "I expect that the preservation of civilization will be based on force, if it is preserved at all. . . ."

LETTER TO HERBERT HOOVER ON "THE WHOLE WORLD SITUATION"

The whole world situation would be much altered if there were an effective defense against bombing by aircraft. There are promising devices, not now being developed to my knowledge, which warrant intense effort. This would be true even if the promise of success were small, and I believe it is certainly not negligible.

In this country there is a great air program, but it consists principally in developing and building aircraft. I have exerted what influence I could

Vannevar Bush to Herbert Hoover, April 10, 1939; Bush to Hoover, April 19, 1943 (Library of Congress, Vannevar Bush Papers, Box 51).

recently toward emphasis of the long-range research aspects of this matter, but with no success. In the whole program, anti-aircraft seems to be almost completely overlooked. The Army and Navy have some development going on, but not nearly enough in my opinion to examine into the various possibilities for progress along these lines. The work is distributed to several branches, and anti-aircraft is no one individual's special concern. There is no centralizing agency on research such as the NACA [National Advisory Committee for Aeronautics] supplies for aircraft research. When military men are queried, they usually reply that "the answer to a plane is another plane." Even if this were true today, it need not be in a few years. The real reason for lack of intense activity lies in the fact that anti-aircraft matters are deeply buried in bureaus or corps which are primarily interested in something else.

Yet there are decidedly interesting new methods available. For one thing the new sources of ultra-high-frequency radiation developed right there at Stanford open up new possibilities of radio detection of planes. I have recently aided in furthering research on this subject at MIT, which has cooperated closely with Stanford. The anti-aircraft aspect of this development is being thought about, but not worked on as far as I know. Yet it offers a very real possibility, in my opinion, of precise and rapid control of guns, even although the target is high and above clouds. There are also other methods warranting attention.

Recently, I sufficiently informed one Army man so that a report on this subject has gone to the head of the War Plans Division. It recommends a new laboratory on anti-aircraft research, joining Army and Navy research with civilian cooperation. I am not sanguine that much will happen as a result.

Having watched the way Congress has handled somewhat similar matters recently, I am pessimistic as to what can be accomplished in that direction. Of course if there were a popular clamor for protection against bombing, it might be different. It might even be possible to produce a popular interest for some of the magazines, and news chains could make quite a story if they were cognizant of the facts, even those already available publicly. However, it seems probable that if such a clamor were aroused it would probably result in distorted emphasis on immediate matters rather than on sound research and development.

Of course this is not a matter for foundations, and my interest is entirely personal. If there is any hope of preventing bombers from crossing important national boundaries, however, I would like to contribute to that result, and I believe that in the long run it can be accomplished. If there is something I can do in a personal way, which will not be an improper thing for the President of the Carnegie Institution to be doing, I would like to get at it. Perhaps the telling of the right story to the right people would be sufficient, but at the present juncture it is hard to see who these [people] might be. The story is known now of course, at least in part, by various technical officers in the [military] services, but it doesn't seem to be understood by those controlling the programs. It is much more than just a story of a possible new detection device; rather I think it is the broad story of a reasonable balance of emphasis between development of aircraft and of ground defense. The Army undoubtedly think they have this now, and would probably resent or disregard a civilian scientist's contrary opinion. I certainly have no intention of breaking into print on the subject. A false move would be apt to do more harm than good.

13

LETTER TO ARCHIBALD MACLEISH ON "ADEQUATE HANDLING OF LARGE MASSES OF PHOTOGRAPHS" (1940)

Bush gave substantial thought in the 1930s to the emerging problem of managing the growing mass of recorded information and to the need to rethink how libraries organized books and served both scholars and citizens. Bringing an engineering mentality to the problem of "information overload," Bush emphasized the importance of rapid retrieval of information and the reliance on codes and coding to store, organize, and retrieve specific pieces of information. For this purpose, he envisioned a device small enough to sit atop a desk that he called a "rapid selector." In MacLeish, a well-known poet and dramatist, Bush found an eager audience for his ideas. Having been asked by President Roosevelt to help modernize the venerable Library of Congress, MacLeish was making the rounds of big thinkers in Washington on the subject of libraries and the organization of information. After a lunch meeting with MacLeish, Bush summarized their discussion and penned a letter that displayed a rare combination of stylish explanation with a visionary picture of humanity's information future.

LETTER TO ARCHIBALD MACLEISH

Dear MacLeish:

We discussed this noon the possible application of the rapid selector to a large file of photographs, and this letter will summarize some aspects of that matter as a basis for further consideration.

The rapid selector is a device, now being perfected at MIT, which very rapidly reviews items on a roll of film, and selects out desired items in accordance with a code. The items are entered in succession photographically onto the roll of film, each with an identifying code in a dot pattern adjacent to it. In order to select items from a film, the roll is placed in the machine, a set of indices are placed in accordance with the code of the desired items, and the film is run through rapidly. Speeds of 1000 items per second can be attained. Every item corresponding to the code will be automatically photographed out onto a strip of sensitive film, which is then developed and ready for use. One limitation occurs as follows: If items to be selected occur nearer together on the original film than ten items apart, the second one will be missed if the speed is at the rate of 1000 per second. With a speed of 100 items per second, none will be missed. At full speed the machine can be arranged to give a signal if perchance it misses an item, whereupon that part of the film can be run slowly to pick it up. Difficulty from this limitation can apparently be avoided by some care in entering the original material.

It would be readily possible to place photographs on the original film as the actual items, each with its own code pattern. The strip of film resulting from selection could then be examined in a reading machine and the selected photographs examined full size. If the ratio of reduction was excessive, some detail might be lost in the process, but the detail would be reasonably present in the film of selected items. If it were desired to make the reduction small, and to have the film of selected items preserve fine detail with great faithfulness, it might be desirable to run the machine relatively slowly, say at 100 pictures per second.

In entering the original material onto the original film, a photograph would be placed in position on a platen and the depression of a lever

Vannevar Bush to Archibald MacLeish, May 8, 1940 (Vannevar Bush Papers, MIT Archives, AC4/42/Bush-1940).

would take the picture and insert it. At the same time the corresponding code would be entered by means of a keyboard.

As presently constructed, the coding area of the machine provides for 12 alphabets. It is of course immaterial whether alphabetical or numerical entry [is] used. As an example, three of these alphabets might be assigned to the geography of the photograph, the first letter giving the country, the second [a] subdivision and the third a close location. Similarly, two more of the alphabets might be assigned to chronology. I give these as examples, although an entirely different method of entering the material might be utilized. The point is that there seems to be plenty of room for fine classification.

In setting up the indices for selection, any portion of the data can of course be used. If one set [has] no indices at all, he would re-photograph out the entire film, although of course this would hardly be done. If one index only were set in the geographical field, then every American photograph might be selected. Similarly if one index in the chronological field were set, then every photograph taken in 1890 would be selected. These, however, are all cases of much too broad generalization. Actually indices would be set finely in some one or two fields and in many cases only a single photograph would be thus sought and selected. The point is that on entering for a search, entire fields may be ignored. Thus one may search with a fine designation geographically and a fine subject designation, paying no attention to chronology and so on. It is important that this feature, which differs radically from the usual indexing scheme, be grasped in approaching the use of the device. Incidentally the machine ordinarily would be set so that, if an unduly broad classification were called for, only say ten pictures would be photographed out, whereupon the machine would stop.

It will be exceedingly interesting to find out whether the problem of adequate handling of large masses of photographs can be approached usefully through this particular instrumentality. I know that the group at the Massachusetts Institute of Technology would cooperate very effectively in any effort to bring this device into serviceable use in a scholarly field, for it is with this primarily in mind that they have carried out the entire development.

Cordially yours,

V. Bush

14

"LEAVE NO STONES UNTURNED IN RESEARCH" (1940)

Bush's first significant statement on technological change and military power came during Germany's invasion of France in May 1940. His brief, forceful message resonated with enthusiasts of air power and came at a time when Germany's aviation technology was the envy of the world. As strength in digital technology is one critical metric of national power today, in 1940, air power provided one measure of military strength. On the eve of World War II, the United States lagged in many areas of technology, but weaknesses in aviation were stark: American planes were much less effective in combat than German planes. Bush implored Americans to support an effort to catch up quickly, and to do so through intensive research in relevant fields. He believed that with a sense of urgency and clear purpose, the United States could accelerate advances in aviation. For the rest of his life, Bush would generalize from specific cases of innovation and argue that technological capacities matter greatly in the rise and fall of great nations. To sustain national vitality, mastery of emerging technologies and new scientific knowledge is crucial, and "constant research" requires steady government support. Bush here sets the table to advocate more forcefully than any other American during World War II for vast public funding of science and engineering and an abiding commitment to pursuing every potential path to new and improved weapons, declaring in an urgent and aggressive tone,

"For war or for peace, we must leave no stones unturned in research."
And perhaps Bush's heirs would add today: And so it remains..

"LEAVE NO STONES UNTURNED IN RESEARCH"

The events of the past few days and weeks have yielded every indication that air power may be the controlling factor in modern war. Strong armies and strong navies are still of primary importance in warfare, but it seems obvious that the nation that can control the air will be dominant.

The fate of a nation, therefore, may depend on its possession of airplanes of superior performance. Superior performance can only be achieved by aircraft designers who are absolutely up to date on their technical information. It is the 'know how' that counts, and 'know how' in any field, and particularly in aeronautics, is acquired only through constant research.

That there should be some direct relationship between military success and the *quality* of research in aeronautics is fairly obvious. Accurate information and a correct interpretation of research results are necessary. But *quantity* of research is also highly important. A large capacity to do research work is as essential as a large capacity to produce aircraft. It would be foolish to create a bottle neck for quality production because of a lack of ability to obtain research results quickly. Under wartime pressures much may depend upon the ability of airplane designers to get correct answers to new problems quickly.

Adding it all together, it is safe to say that the course of history is largely being influenced by scientific investigations in aeronautical research laboratories. The outcome of the aerial battles of tomorrow are being decided today by the men who are working in wind tunnels, towing tanks and engine laboratories.

Vannevar Bush, excerpt from an address to the National Aviation forum, entitled, "Aeronautical Research, a Vital Link in Our National Defense, May 29, 1940" (NASA, History Office).

15

"TO THE THINGS OF THE MIND" (1941)

Memorandum Regarding Memex

Bush long pondered how technology could automate human thought and how what the industrial revolution did for manual labor, the coming revolution in information storage and retrieval might bring to individual human creativity and to the collective knowledge and "intelligence" of humanity. In this selection from April 1941, where Bush presents his fullest and earliest detailed account of his conception of automated information storage and retrieval, we can identify the origins of Bush's radical vision of the future of personal computing, the organization of stored information, and the new horizons for human creativity in the coming age of rapid scientific advance. The cover letter conveys Bush's skepticism and humility about the potential for the automation of human thought. Writing to Eric Hodgins, an editor at Fortune *(a prominent magazine and a sibling publication of Henry Luce's* Time *magazine), Bush seeks editorial guidance and advice on how to present his message to wider audiences about the information revolution. Over the ensuing years, Bush and Hodgins would become close, and Hodgins would assist Bush in writing articles for popular magazines and his autobiography,* Pieces of the Action, *published in 1970. In this memo, Bush presents core ideas that would animate his landmark "As We May Think" article, published in the* Atlantic *magazine in July 1945.*

When it came to predicting the future of technology, Bush sometimes erred. The memex was not one of those times. Subsequent generations of computer scientists and engineers credit him with enormous prescience

about the course of information technology, especially in the way infor-
mation is organized, stored, and retrieved via human-built machines. The
memo also offers a snapshot of Bush's private views on the state of science
and society relations on the eve of U.S. entry into World War II. Parts of
the essay also anticipate Bush's introduction to his 1945 report Science, the
Endless Frontier.

"TO THE THINGS OF THE MIND": MEMORANDUM REGARDING MEMEX

April 10, 1941
Dear Eric:
I have written out, and I enclose, the memorandum I spoke of.

This attempts to summarize why [I wrote an article about the] Memex was written and what it was hoped it might accomplish. Then it goes on to recite how times have changed, and why the original idea is not now likely to arrive in the form initially constructed.

It [the memo] ends by stating that I don't see just what to do about it.

Perhaps the idea behind the thing has just evaporated and left merely a story with no great excuse for telling it.

Yours,
V. Bush

"TO THE THINGS OF THE MIND"

The general objective in writing the article was to influence in some small way the thinking regarding science in the modern world, and to do so in an interesting manner. The secondary objective was to emphasize the opportunity for the application of science in a field which is largely neglected by organized science.

Vannevar Bush to Eric Hodgins, April 10, 1941 (Vannevar Bush Papers, Library of Congress, Box 50, Eric Hodgins File, General Correspondence).

The audience aimed at was the group of active thinkers who are taking part in affairs, rather than those who are developing a philosophy concerning the relation of science to human progress or the like. Hence the decidedly objective approach. In particular it seemed worthwhile to try to influence the thinking of trustees of foundations, educational and scientific institutions, and those concerned with governmental support or control of scientific efforts.

The concern was primarily with the physical sciences, rather than the natural sciences as a whole, for the conflict as to what constitutes a reasonable attitude centers principally on physics and chemistry and their engineering applications.

The main thesis is that science and its applications are not, on the whole, evil. A secondary thesis is that gadgetry is not necessarily trivial, and that in particular [gadgets] may contribute substantially to man's mental development in the future as it has done in the past.

The attitude of the intended audience has gone through much transformation during the previous two decades.

There was much naive praise of the benefits of science in all sorts of fields up to the last war. After the war this [pro-science] attitude became intensified, and applying science in the American way was about to solve all the ills of mankind, increase the standard of living around the world and leave us all prosperous and happy.

This was [animated] by the idealism that abounded. All war was over. We could leave these evil uses we had indulged in and turn to higher things, such as air-conditioning and four-wheel brakes.

Then came a disillusionment and a reaction. The grand economic idea came a cropper. The reaction extended to the applications of science; we had indulged in too much change in our ways of life and should become simple. Technocracy convinced many people that the machines meant unemployment.

The idealism persisted in a new form. If it were evil to devise new material things for ordinary use, it was still more evil to make new devices for war. We should show a fine example and forget all such things. We nearly did.

The effect was principally with older men, and with laymen. Youngsters still tried to become aeronautical engineers, and physicists were as keen on cracking the atom as ever. Foundations, however, radically cut their support of the physical sciences at this time. They never had participated much in the applications of physical science, leaving that for those having a commercial motive, and now they abandoned this field entirely. This was true even though there were plenty of possible applications which had little commercial incentive attached, but which were nevertheless of much potential benefit. For example, the foundation never thought of doing a fine job, beyond that which would be caused by a profit motive, in producing good mechanical music. It was left out because it was mechanical. Rather the foundations turned to the biological sciences, which could satisfy man's craving for knowledge without doing him any harm, and in particular to the medical sciences which could cure his ills, sometimes without inquiring too deeply into the ultimate results. They also turned to the social sciences, often with a hope that was entirely unjustified by the state of advancement of the disciplines in that category. They turned heavily to art, to adult education and so on. Many new foundations were created, with very great sums in the aggregate at their disposal, and almost uniformly they avoided the physical sciences and their applications. The attitude foundations exemplified a more general attitude which tended to frown upon physicists and all their works.

This was the situation when the Memex [article] was first written; and this was the attitude which it was intended indirectly to combat. Publication was delayed, principally because it took a long time to write.

Then came the present war and attitudes changed in a hurry. Where there had been a feeling that we had too much physics, there arose the cry for more and very active physicists to devise war gadgets. The shift in attitude has not been made by all. Most men on the active list have shifted, if they ever needed to. Many a philosopher hasn't, and of course many thoughtful people do not know there is a war on. But the change of attitude has been profound.

The idealism has largely gone. We are living in a real and tough world. It is no longer regarded as wicked to devise a means for shooting an airplane, in a world where it suddenly appears that men will actually ride in airplanes and drop bombs. It is not even wicked to work out more powerful bombs to drop on someone else. It is being realized with a thud that the world is probably going to be ruled by those who know how, in the fullest sense, to apply science, whatever their other attributes may be.

Part of the original thesis no longer needs demonstration. It may now be all the more necessary to emphasize the rest.

The long-range pursuit of scientific knowledge is now being pressed aside by the needs for defense research. This is as it should be, but it will be highly important to swing back again, when the pace is not so breathless, to the sound development of fundamental science and the widespread acquaintanceship with scientific matters which goes with this. There is real danger, when peace comes, if we still have an orderly civilization that is controlled by mass public opinion, that we will again lapse into dreamland.

The application of science to war is now getting its emphasis. Along with this, and partly as a consequence after the emergency, it is more probable that there will come an increased emphasis, at least for a time, on applications to transportation, food, clothing and all the other aspects of living in a mechanical way. It may therefore still be pertinent to argue that there are possible applications to the things of the mind that are not negligible.

It is a question of timing. People will even now read an interesting story. It may even carry a new realization that there is a neglected field worth cultivating. But the field cannot be cultivated now, and by the time it can, the story, if published now, will undoubtedly have been forgotten. Even so, it may now be cheering to contemplate the application of science to something more satisfying than weapons. Perhaps it is worth telling for that reason. It was originally constructed to produce an effect, and the need to do so still remains. It seems too bad to tell it at a time when it can probably accomplish little of its original purpose. Yet if merely allowed to wait it probably will not last until it could be told in an atmosphere which would admit its full effectiveness, whatever that might turn out to be.

It is certain, however, that if [my article about the future of individual devices to automate thought and knowledge] is to be anything more than an amusing take, it needs to be revamped to bring it in line with changed times. Just how to do this is a puzzle. Another thing to do would be to say, "when this was written . . ." and then tell a yarn for the sake of telling a yarn, which is not a very serious undertaking. My [aim] to influence the attitude toward science would then have to be approached from an entirely new direction when and if it were possible to do so.

16

SCIENCE AND NATIONAL DEFENSE (1941)

Writing forty-five days before the Japanese attack on Pearl Harbor, and the declaration of war on the United States by Germany, Bush gives his first detailed description of the U.S. mobilization of science, engineering, and innovation in the service national security and preparations for war. First delivered as a speech and then republished in a leading academic physics journal, "Science and National Defense" was intended for an audience of academic and industrial scientists and engineers who Bush knew were wary, or even hostile, to working closely with military officers and government officials. Bush sought to build conceptual and practical bridges between key actors in the domains of technology, science, and the nation-state. While he praised the talents of researchers, he emphasized the value of close coordination between soldiers and civilians and, through a blizzard of acronyms and shifting organizational forms, advanced the overall point that sound planning, management, and accountability were more decisive to winning wars than sheer ingenuity. For Bush, independence was crucial, not as an ideological principle or abstract belief, but rather because critical thinking and free exchange of ideas promoted beneficial feedback, thus allowing new weapons, and new processes, to be quickly refined and revised in order to suit changing and unexpected challenges. In short, Bush conveyed a message he would repeat in the postwar era: research advances are only part of the answer; strong management, planning, and organization usually matter as much or more.

SCIENCE AND NATIONAL DEFENSE

In this discussion of the present position of science and research in national defense I will confine myself to two points. The first concerns the form of organization under which the scientists of this country are working. The second, which is very brief, has to do with the spirit with which the task is undertaken. As to the work itself I cannot, of course, be specific at the present time.

Details of the organization have been made known, but I think they are not well understood generally. In June 1940 there was formed, by order of the Council of National Defense, a group called the National Defense Research Committee (NDRC), for the purpose of supplementing the work of the Army and Navy in the development of devices and instrumentalities of war. This new organization was intended to function in an executive, not an advisory, capacity. The advisory function was being adequately cared for by the National Academy of Sciences, which has been in existence since the Civil War period, having been created by Act of Congress for the express purpose of advising Government on its scientific and technical problems. There was, however, need for a civilian group with executive powers to supplement the scientific and technical work of the Army and Navy, for, in an amergency, an expansion of the scientific attack on war problems was essential, and the Army and Navy could not themselves undertake this immediately and fully. This need resulted in the formation of the NDRC as a means for enrolling a large number of civilian scientists to assist in research on problems of national defense as promptly as it could be done. The NDRC was formed as an operating part of the emergency governmental machinery, in contrast with the position of the National Academy of Sciences, which is an independent organization, operating under a Congressional charter which defines its obligation to render advice when called upon by government agencies. The NDRC thus is intended as an emergency organization and is not of permanent character as is the Academy.

Bush addressed his words to the Joint Luncheon of the Acoustical Society of America, the Optical Society of America, and the Society of Rheology (dedicated to the science of "deformation" and the flow of matter) in New York on October 24, 1941. The text was published in the *Journal of Applied Physics*, December 1941.

Let me say in passing that, in order that the Academy and the Research Council may most effectively carry on their exceedingly important work, it is very essential that they maintain their independent status, for this gives added force to their opinions on scientific problems as expressed to agencies of government.

NDRC consists of six civilians (including the President of the National Academy of Sciences and the Commissioner of Patents), an officer of the Army, and an officer of the Navy. Initially it was organized in four divisions: one under R. C. Tolman, of the California Institute of Technology; one under K. T. Compton, President of the Massachusetts Institute of Technology; one under F. B. Jewett, President of the National Academy of Sciences; and one under J. B. Conant, President of Harvard University. There are now about 60 sections in these divisions, composed of voluntary part-time and full-time workers, plus a few technical aides who are paid by the government.

The Committee operates primarily by means of contracts with universities, colleges, research institutes, and industrial laboratories. These contracts are initiated and supervised by the various sections and there are now about 450 of them in operation with nearly 120 contractors. These contracts are drawn with the intent that the contractor, whether university or industrial laboratory, shall neither gain nor lose financially through participation in these defense research activities. Such an aim is difficult to achieve, of course, but we are now making a careful study of the situation to see how close we have come to this goal.

The NDRC has tried to carry out its work with a minimum of interruption to the regular affairs of the universities. Of course it is not possible to proceed without disruption and inconvenience, but that has been held down as far as possible. Many a physics department throughout this country has nevertheless been put to very great stress. The men who have carried on under a heavy overload, continuing the work of the university in order that some of their colleagues might participate directly in the defense program, are contributing no less to the national interest than are those immediately engaged in the sections of NDRC or otherwise.

The number of people involved in the work of the Committee is rapidly increasing. There are now about 500 individuals in the NDRC organization who serve as members of sections, consultants, and so on. About 2000 scientific men are engaged in defense research in connection with NDRC contracts, and probably there is an equal number of helpers.

The NDRC spent about ten million dollars during its first year, and it had nearly as much for the second year, beginning July 1, 1941. If the President signs the bill which was passed by the Senate yesterday, another large sum will become available. I feel quite sure that if a bottleneck should develop, it would not be caused by the number of problems that are important and should be worked on, nor by lack of funds with which to carry on the work, but by a shortage of available personnel. For example, a recent study indicated that, of the available physicists whose names are starred in *American Men of Science*, about 75 percent are now engaged in war research in one way or another. I think the 25 percent who are still available will soon be called upon. The call for chemists has not been so great as it was in the last war; about 50 percent is the corresponding figure. In addition, many engineers are thus engaged but in their case it is not easy to arrive at a comparable figure.

Effort has been made to spread the work, but that has not been possible in some instances. For some problems it has been necessary to gather an outstanding group in one place in order to provide the advantages of a concentrated attack. Such concentration has occurred under several universities in this country, men having been brought from other institutions for that purpose. Within this next year, however, we hope to spread the work much more than was practicable in the great haste of initiating activities in the summer of 1940.

The special function of NDRC, as previously stated, is to supplement the work of the Army and Navy in the development of devices and instrumentalities of war. Let me say immediately that, in spite of difficulties due to lack of funds in the years preceding this emergency, the Army and Navy, in my opinion, have done a very fine piece of work in research, in development, and in proceeding toward advanced instruments of war. We are supplementing previous work, not starting anew, in almost every field in which we operate.

There is very close liaison with the Army and Navy, each section of NDRC having its own liaison officers. These officers represent some of the brightest, keenest men in the armed services on the technical front. Relations have been cordial, and I believe that many a scientist has gained, through his work with the Army and Navy, a new respect for the burden being carried and for the quality of the men who are carrying it. I think, too, that many an officer in the Army and in the Navy has gotten a somewhat different idea, of course, in the past year, of the scientists of the country. They have found that the scientist is not necessarily a person with "long hair," but that he can attack a problem in a practical way, and that he can work long hours and take it with the best. As a result, there has developed a mutual respect which is very heartening.

Having thus given you the salient points in regard to NDRC, I will now tell you of a new organization. Last June, after just one year of operation by the NDRC, the President, by Executive Order, established the Office of Scientific Research and Development (OSRD), of which I am Director, to coordinate all defense research wherever it might occur. The OSRD has two major divisions, one being the National Defense Research Committee, which continues as before, except that Dr. Conant succeeded me as Chairman and Dr. Roger Adams has taken over Dr. Conant's previous duties. In addition, there is a business office under Dr. Stewart and a liaison office under Mr. Wilson, the latter being principally engaged in handling our relations with the British.

The second major division of OSRD is the newly formed Committee on Medical Research, of which Dr. [Alfred Newton] Richards, of Pennsylvania, is Chairman. This Committee is constituted in much the same way as the NDRC, and it shares with NDRC the funds furnished by OSRD, in order to conduct medical defense research. It works primarily with the Division of Medical Sciences of the National Research Council, which had already been active in this field and had organized committees necessary to carryon such work, of which there are about 50. The Committee on Medical Research has close relations with the U. S. Public Health Service, as well as with the Surgeons General of the Army and Navy, having in its membership a representative from each of the three offices.

. . . . The OSRD is aided by many organizations that are not directly concerned in the activities of its two main groups. I mention, for example, the National Roster of Scientific and Specialized Personnel under Dr. Carmichael, which, together with the National Research Council, is undertaking the very important task of locating competent personnel. There is also the National Inventors Council, formed under the Department of Commerce for the purpose of evaluating the very large number of suggestions coming from the public. The Inventors Council has a difficult and somewhat thankless task, but is doing it very well indeed. The wheat they sift from the chaff is passed on to the Army and Navy for development as needs arise.

Thus there exists in the organization of [OSRD] the basis for an effort commensurate with the importance of science in modern war, which is a very high order of importance indeed. Of the work itself I cannot of course tell at this time, but it will be a striking story when it is finally revealed. With 60 active sections in the NDRC and 50 committees in the CMR, the range of work is obviously large. I can, however, mention one thing that has already been made public. When the Battle of Britain took place in August, September, and October 1940, [the German] invasion failed for two reasons: First, the British fighting forces in the air were courageous, skillful, and well-equipped, better equipped than were the invaders in many ways; secondly, radio detection, developed by a group of devoted British scientists working from 1935 on, at times without much encouragement, offset the element of surprise. This one development may have saved the isle of Britain. It is one field of obvious importance; others undoubtedly occur to you.

Most of the matters that OSRD handles are quite naturally clothed with the mantle of secrecy, for every precaution must be taken in dealing with military matters of great potential importance. The various sections are working in specific fields, and the affairs of the organization have been compartmentalized in order better to follow the general policy of permitting a man to learn confidential things only to the extent that is necessary in order that he may function effectively. Another rule is that, in working with people outside the organization, OSRD members listen and do not talk. It is not the most agreeable way of working; it is not natural for a scientist, but it is necessary under present circumstances. Appointments to posts in the organization are made only when the Army and Navy, after careful investigation, have indicated that they have no reason to suspect that there is not complete loyalty.

I said that there were two points to be considered. One is the organization, but organization is very little. The spirit in which the work is conducted is much more important. There is no unanimity in this country as to how or when or where or to what extent the power of this nation should be exerted to defend our way of life. But there is unanimity on the thesis that the power of this country must be increased at once and to the maximum possible extent.

The scientists of this country have done more than speak on this subject. They have taken off their coats and gone to work, and much academic research has been postponed in the process. The matter of credit has been utterly forgotten. They have shown a willingness to work under necessarily rigid restrictions, as well as with a reasonable tolerance of the petty inconveniences and annoyances that are inevitable in the confusion of adapting themselves to military ways. They have shown that they are willing to go into a strange ballpark and learn the local ground rules. In only a year they have done things. Ordinarily, it is at least three years from an idea in the laboratory to its use, and yet I say to you that results are being obtained, and they are taking form in copper and iron.

Those of you who are privileged to participate in this work, as I am, will find therein a deep satisfaction, even though it substitutes for a thing we held more highly: the privilege of contributing to the growing knowledge of the race. Those of you who are not participating directly, but are holding the fort in order that your colleagues may participate, or who are carrying on in a field where the thread of growing knowledge might otherwise be broken in the present distress of the world, will also look back some day to this period, not only as a time of stress, but as a time when we were all privileged to participate in one thing on which we could become united: the defense of the country to which we owe our allegiance. The scientists of this country are united, and they are obtaining results.

17

EDISON AND OUR TRADITION OF
OPPORTUNITY (1944)

Bush engineered the largest and most complicated research and discovery in human history during World War II. He drew on the talents of scientists and engineers all over the world, but his management style depended heavily on enlisting a handful of elite universities, especially MIT and Johns Hopkins, and major industrial corporations, notably DuPont, to deliver usable innovations that could immediately influence the outcome of battles and the overall war with Germany and Japan. While promoting the concentration of research and development capacity, and while creating structures that embedded even the most independent-minded scientists and engineers in secretive bureaucracies, Bush maintained in public that the era of the pathbreaking individual inventor and daring technological entrepreneur was not over. Bush insisted—in the face of what seemed like rising corporate and government domination of innovation—that future Edisons were not only possible but inevitable and necessary. Speaking before a gathering of electrical engineers in 1944, Bush declared: "There must remain in the United States the opportunity for an Edison, the opportunity for any youth with initiative, resourcefulness, practicality and vision, to create in his own name and by his own efforts new things that will tend to make this country vigorous and strong and safe."

EDISON AND OUR TRADITION OF OPPORTUNITY

Thomas A. Edison exemplified an important aspect of American life. He combined, in a superlative degree, resourcefulness and initiative with an intense practicality and a keen vision. The many results which he attained during his life were the blossoming of this combination of talent in an environment which was wonderfully adapted for such a growth.

I do not need to say that these are not the only characteristics that are needed in scientific and engineering endeavor. The list of Edison medalists is itself a demonstration of the recognition of the need for combining such qualities with scientific deduction, mathematical analysis, and the like, in order that there may be a rounded whole. Our great industrial advance, with its intense use of electric power and its mechanization, has risen by reason of the efforts of many types of mind. But without the important phase that Edison so notably exemplified, the scene would indeed be a drab one.

The United States has prospered in a material way. It has attained a standard of living far beyond that arrived at in any other country. This is not entirely explained simply by the presence of great resources, or a large and uniform market, or even by the opportunities for pioneering that are summarized in the convenient term 'the frontier.' Neither does the opportunity for advancement necessarily vanish when the geographical frontier gives place to one that is entirely of a technical nature. The advancement has occurred because America brought a new idea into the world: a valuable idea.

In a word, there was produced an atmosphere in which Edison could function, and in which men like him having ideas and the intellect for their furtherance, however much they might differ from him as to method, could make their influence felt. The startling forward steps that Edison caused by his own efforts could not have occurred in a totalitarian state, whether the labels were that of state socialism or Fascism or something similar. They could occur fully only under the unusual state of circumstances which obtained in this country when Edison worked and

---

Bush gave this address to the American Institute of Electrical Engineers on January 26, 1944.

created. The atmosphere that existed is hard to define; it had its crudities, but it was a unique atmosphere and it is well worth preserving.

As the United States matures, we are in distinct danger of losing this enormous asset. The unwillingness to take risks which accompanies economic maturity, the crosscurrents of pressure interests, the mere increase of interdependence due to the advance of mechanization, all tend to destroy the flexibility and freedom of action which are essential parts of the atmosphere of individual creation.

To the trends that were molding our environment and conditioning the atmosphere within which our pioneers worked was suddenly added the impact of the war. We awoke late from a period of lethargy to find ourselves in an intense struggle caused by the clash of our political and social philosophies with ideologies far different from our own. We shall not return to exact condition from which we departed when war came. We have learned many things, and it is to be hoped that we shall not forget them. It is of especial importance that, as we re-adapt ourselves to a new period of peace, we should preserve the essential features which made this country great. In particular, the opportunity for the individual creator, for the industrial pioneer, for the inventor with vision and practicality, should not be lost; and the atmosphere for our industrial life should be favorable for his efforts on our behalf.

The effort to hold our own in a technical contest was not by any means a sinecure, or a struggle in which the outcome could be predicted in *a priori* and with full confidence. In the course of the war with the Axis nations, we [have] had to mobilize our technical and scientific resources to the full, and we [have] had to do so with the confusion that is inevitable when a great lumbering democracy suddenly turns to war. It was necessary to regulate, to impose controls, and to operate with a secrecy that is foreign to our own usual open methods. It was not necessary, however, to suppress thought and initiative and the democratic functioning of science and technical groups devoted to important phases of the technical aspects of the war effort. The results were good.

When the full tale is told, or the part of it is told which properly can be realized, I am convinced that it will show one thing and that it will show this forcefully. The free untrammeled science and engineering of a democracy, when it once becomes directed toward an objective with full vigor, can outstrip the regimented efforts of any totalitarian state, provided there is anything approaching equality of resources on the two sides. I believe that it will come to be realized fully that our progress in manufacturing with enormous speed the weapons needed by our Armed Forces, in improving these weapons, in initiating new ones, and in keeping our civilian economy with its enormous technical complication still running effectively, was not matched by the enemy [in World War II], and that this occurred in no small degree because our wartime effort was built upon the existing structure of freedom, with all that the word implies.

This [structure of freedom] must be preserved. As we turn again to the days of peace, it must be preserved in spite of the trends that flow from greater interdependence of society, in spite of the trends that are inherent in the growing maturity of a nation, in spite of selfish interests of every sort. There must remain in the United States the opportunity for an Edison, the opportunity for any youth with initiative, resourcefulness, practicality and vision, to create in his own name and by his own efforts new things that will tend to make this country vigorous and strong and safe.

The way in which this can be accomplished is not immediately apparent. The difficult does not reside in the attitude of the people, for certainly there is an overwhelming majority that holds strongly to the conviction that freedom of individual opportunity must not be allowed to lapse, and that this involves genuine industrial opportunity for the individual, or the small group of individuals that join in a business effort. The difficulty arises from two facts. First there are great sections of our industrial affairs that can be handled economically only by large industrial units. Second the very complexity of modern life requires increased and centralized governmental activity, in order that the public interest may be fully protected and furthered by those measures of regulation and public works which government alone can perform adequately. The point is that, as a people, we have two parallel objectives, and we have been clumsy at times and allowed one to submerge the other.

In time of war concentration of effort and the imposition of rigid controls are essential for a fully coordinated all-out effort. The question is raised why, if this concentration is more effective for war, is it not also more effective for peace?

One answer is the fact that the war concentration is effective primarily because it was constructed out of elements that became available and efficient under relative freedom of peacetime conditions in a democracy, and there is no assurance that a brief effective concentration would remain so. Quite the contrary is what I believe the record of history will show. A more complete reply lies in the fact that we seek in the United States something beyond mere mechanical efficiency. We seek a society in which initiative and talent may have an outlet, and in which the individual may have opportunity to rise by his own efforts and contributions and not merely by the fixed operation of a system. We would be willing to sacrifice, if need be, some mechanical efficiency in the interests of individual freedom. My own conviction is that no sacrifice in the full effectiveness of the country is truly involved in the long run; and that, on the contrary, the elements which have rendered us strong in the past will render us stronger still in the future if we have sufficient intelligence and conviction to insist upon their preservation.

As peace returns, controls automatically become relaxed. This is not enough. A negative slump back to a state of drifting, buffered by the trends that are inherent in technical progress, will not carry us on our appointed course as a nation. A positive and well-thought-out effort is needed if we are to combine the conditions of modern industry and modern government with that freedom within unitary organizations and within society as a whole which will allow the unusual individual to have a real chance of accomplishment and success. It can be done, if we have the steadfastness and the conviction to insist that it be done. In this effort the great body of engineers and scientific men in the United States have an important part, for they are in a unique position to view and understand both sides of the matter which is primarily technical, economic, and organizational. If we are wise there will be, in the future, many Edison's in the United States. They may not shine with his peculiar brilliance, but they will add to the well-being of the nation a necessary element which can be added in no other way. Their opportunity must be preserved open before them.

18

SALIENT POINTS CONCERNING FUTURE OF
ATOMIC BOMBS (1944)

Along with James Conant, Bush was the first American to warn a U.S. presi-
dent of the dangers of nuclear weapons and raise the possibility that another
country would build an atomic bomb in the near future. He cowrote this
warning with Conant, the president of Harvard; Conant also was Bush's
deputy during World War II. The two men had much to do with the organi-
zation and management of the Manhattan Project. Because of Conant's sta-
tus as an eminent chemist and leader of Harvard, and Bush's close personal
relationship with President Roosevelt and with the Secretary of War, Henry
Stimson, this scientific duo was uniquely positioned to study and reflect
on the world's nuclear future. Significantly, nearly a year before the atomic
explosion over Japan, Bush and Conant accurately predicted that other
countries also would create atomic weapons in the foreseeable future. They
warned of the need to establish an international control system to prevent
a postwar nuclear arms race that could end in the destruction of human
civilization. For Bush, anxieties over nuclear weapons and ideas about how
to manage and control the destabilizing technologies that produced these
superweapons persisted long after the end of World War II in 1945. Though
Bush would lose his privileged status within the nuclear priesthood by the
early 1950s, nuclear weapons cast a long shadow over his life, and his think-
ing on how to avoid nuclear war coevolved and coexisted with his creative
ideas about a range of less existential subjects, including computation and
the relationship of science to government, for the rest of his life.

SALIENT POINTS CONCERNING FUTURE OF ATOMIC BOMBS

Office of Scientific Research and Development
1530 P. Street N.W., Washington DC
September 30, 1944
The Secretary of War
Washington, DC

Dear Mr. Secretary:

In response to your suggestion, we outline in the two attached memoranda our thoughts on the international postwar aspects of the special projects. We believe that the following points are correct, and of great importance to the future peace of the world.

1. By next summer this will become a matter of great military importance.
2. The art will expand rapidly after the war, and the military aspects may become overwhelming.
3. This country has a temporary advantage which may disappear, or even reverse, if there is a secret arms race on this subject.
4. Basic knowledge of the matter is widespread and it would be foolhardy to attempt to maintain our security by preserving secrecy.
5. Controlling supplies of materials cannot be depended upon to control use, especially in the forms which the subject may take in the future.
6. There is hope that an arms race on this basis can be prevented, and even that the future peace of the world may be furthered, by complete international scientific and technical interchange on this subject, backed up by an international commission acting under an association of nations and having the authority to inspect.

Very sincerely yours,
V. Bush
J.B. Conant

National Archives (U.S.), Bush-Conant Papers. Only the first of two memos is included. The second memo, while long, restates the six points in greater detail and is not included here.

Memorandum, Sept. 30, 1944
To: The Secretary of War
From: V. Bush and J. B. Conant
Subject: Salient Points Concerning Future International Handling of the Subject of Atomic Bombs

1. *Present Military Potentialities:*

There is every reason to believe that before August 1, 1945, atomic bombs will have been demonstrated and that the type then in production would be the equivalent of 1,000 to 10,000 tons of high explosive in so far as general blast damage is concerned. This means that one B-29 bomber could accomplish with such a bomb the same damage against weak industrial and civilian targets as 100 to 1,000 B-29 bombers.

2. *Future Military Potentialities:*

We are dealing with an expanding art and it is difficult to predict the future. At present we are planning atomic bombs utilizing the energy involved in the fission of the uranium atom. It is believed that such energy can be used as a detonator for setting off the energy which would be involved in the transformation of heavy hydrogen atoms into helium. If this can be done, a factor of a thousand or more would be introduced into the amount of energy released. This means that one such super-super bomb would be equivalent in blast damage to 1,000 raids of 1,000 B-29 Fortresses delivering their load of high explosive on one target. One must consider the possibility of delivering either the bombs at present contemplated or the super-super bomb on an enemy target by means of a robot plane or guided missile. When one considers these possibilities, we see that very great devastation could be caused immediately after the outbreak of hostilities to civilian and industrial centers by an enemy prepared with a relatively few such bombs. That such a situation presents a new challenge to the world is evident.

3. *Present Advantage of the United States and Great Britain is Temporary:*

Unless it develops that Germany is much further along than is now believed, it is probably that present developments in the U.S., undertaken in cooperation with Great Britain, put us in a temporary position of great ascendancy. It would be possible, however, for any nation with good technical and scientific resources to reach our present position in three or four years. Therefore it would be the height of folly for the U.S. and Great

Britain to assume that they will always continue to be superior in this new weapon. Once the distance between ourselves and those who have not yet developed this art is eliminated, the accidents of research could give another country a temporary advantage as great as the one we now enjoy.

4. *Impossibility of maintaining complete secrecy after the war is over:*

In order to accomplish our present gigantic technical and scientific task, it has been necessary to bring a vast number of technical men into the project. Information in regard to various aspects of it is therefore widespread. Furthermore, all the basic facts were known to physicists before the development began. Some outside the project have undoubtedly guessed a great deal of what is going on. Considerable information is already in the hands of various newspaper men who are refraining from writing stories only because of voluntary censorship. In view of this situation it is our strong recommendation that plans be laid for complete disclosure of the history of the development and all but the manufacturing and military details of the bombs as soon as the first bomb has been demonstrated. This demonstration might be over enemy territory or in our own country, with subsequent notice to Japan that the materials would be used against the Japanese mainland unless surrender was forthcoming.

5. *Dangers of partial secrecy and international armament race:*

It is our contention that it would be extremely dangerous for the U.S. and Great Britain to attempt to carry on in complete secrecy further developments of the military applications of this art. If this were done, Russia would undoubtedly proceed in secret along the same lines and so too might certain other countries, including our defeated enemies. We do not believe that over a period of a decade the control of the supply could be counted on to prevent such secret developments in other countries. This is particularly true if the super-super bomb were developed for the supply of heavy hydrogen is essentially unlimited and the rarer materials such as uranium and thorium would only be used as detonators. If a country other than Great Britain and the U.S. developed the super-super bomb first, we should be in a terrifying situation if hostilities should occur. The effect of the public reaction of the uncertainties in regard to the unknown threat of this new nature would be very great.

6. *Proposed international exchange of information:*

In order to meet the unique situation created by the development of this new art, we would propose that free interchange of all scientific information on this subject be established under the auspices of an international office deriving its power from whatever association of nations is developed at the close of the present war. We would propose further that as soon as practical, the technical staff of this office be given free access in all countries not only to the scientific laboratories where such work is contained, but to the military establishments as well. We recognize there will be great resistance to this measure but believe the hazards to the future of the world are sufficiently great to warrant this attempt. If accurate information were available as to the development of these atomic bombs in each country, public opinion would have true information about the status of the armament situation. Under these conditions there is reason to hope that the weapons would never be employed and indeed that the existence of these weapons might decrease the chance of another major war.

J. B. Conant
V. Bush

19

THE BUILDERS (1945)

This brief essay (a mere 750 words) is a praise song to the value of vision and sustained effort by diverse participants, in the construction of knowledge, for the building of an "edifice" of human reflection and insight. Published in the Atlantic *magazine as the war in Europe ended, "The Builders" gives a special shoutout to "those men of rare vision who can grasp well in advance just the [building] block that is needed" in the rising edifice of knowledge. Bush calls these pivotal actors "master workmen."*

As the global war is nearing an end, ordinary people are realizing anew that innovation and scientific advances were decisive factors in the making of history. In this essay, Bush conveys in a rather romantic and literary manner the elusive, mysterious, and cherished process of exploration and discovery in pursuit of knowledge, practical outcomes, or both.

THE BUILDERS

The process by which the boundaries of knowledge are advanced, and the structure of organized science is built, is a complex process indeed.

"The Builders" appeared, most prominently, on single page in the May 1945 issue of the *Atlantic* magazine and was reprinted in a volume of Bush essays, published in 1946 by Public Affairs Press, under the

It corresponds fairly well with the exploitation of a difficult quarry for its building materials and the fitting of these into an edifice; but there are very significant differences. First, the material itself is exceedingly varied, hidden and overlaid with relatively worthless rubble, and the process of uncovering new facts and relationships has some of the attributes of prospecting and exploration rather than of mining or quarrying. Second, the whole effort is highly unorganized. There are no direct orders from architect or quarry master. Individuals and small bands proceed about their businesses unimpeded and uncontrolled, digging where they will, working over their material, and tucking it into place in the edifice.

Finally, the edifice itself has a remarkable property, for its form is predestined by the laws of logic and the nature of human reasoning. It is almost as though it had once existed, and its building blocks had then been scattered, hidden and buried, each with its unique form retained, so that it would fit only in its own peculiar position, and with the concomitant limitation that the blocks cannot be found or recognized until the building of the structure has progressed to the point where their position and form reveal themselves to the discerning eye of the talented worker in the quarry. Parts of the edifice are being used while construction proceeds, by reason of the applications of science, but other parts are merely admired for their beauty and symmetry, and their possible utility is not in question.

In these circumstances, it is not at all strange that the workers sometimes proceed in erratic ways. There are those who are quite content, given a few tools, to dig away unearthing odd blocks, and apparently not caring whether they fit anywhere or not. Unfortunately there are also those who watch carefully until some industrious group digs out a particularly ornamental block, whereupon they fit it in place with much gusto, and bow to the crowd. Some groups do not dig at all, but spend all their time arguing as to the exact arrangement of a cornice or an abutment. Some spend all their days trying to pull down a block or two that a rival has put in place. Some, indeed, neither dig nor argue, but go along with the crowd, scratch here and there, and enjoy the scenery. Some sit by and give advice, and some just sit.

---

title of *Endless Horizons*, and in a different set of essays, also out of print, published in 1967 by William & Morrow Co. under the title *Science Is Not Enough*. The essay originally appeared earlier in 1945 in MIT's *Technology Review*.

On the other hand there are those men of rare vision who can grasp well in advance just the block that is needed for rapid advance on a section of the edifice to be possible, who can tell by some subtle sense where it will be found, and who have an uncanny skill in cleaning away dross and bringing it surely into the light. These are the master workmen. For each of them there can well be many of lesser stature who chip and delve, industriously, but with little grasp of what it is all about, and who nevertheless make the great steps possible.

There are those who can give the structure meaning, who can trace its evolution from early times, and describe the glories that are to be, in ways that inspire those who work and those who enjoy. They bring the inspiration that not all is mere building of monotonous walls, and that there is architecture even though the architect is not seen to guide and order.

There are those who labor to make the utility of the structure real, to cause it to give shelter to the multitude that they may be better protected, and that they may derive health and well-being because of its presence.

And the edifice is not built by the quarrymen and the masons alone. There are those who bring them food during their labors, and cooling drink when the days are warm, who sing to them and place flowers on the little walls that have grown with the years.

There are also the old men, whose days of vigorous building are done, whose eyes are too dim to see the details of the arch or the needed form of its keystone, but who have built a wall here and there and lived long in the edifice; who have learned to love it and who have even grasped a suggestion of its ultimate meaning; and who sit in the shade and encourage the young men.

20

TEAMWORK OF TECHNICIANS (1945)

In reflecting on how the United States achieved unrivaled strength in research, discovery, and invention during World War II, Bush repeatedly emphasized the importance of organization, management, and teamwork (in contrast, say, to citing individual genius). While he understood and accepted the value of individual talent and the unique aspects of American history and society, Bush viewed the mobilization of talent and resources, including but not limited to science and engineering, as ultimately a triumph of planning and administration. While the cult of the individual looms large in American mythology, Bush chose instead to identify management and leadership as the crucial ingredients in the achievement of wartime supremacy. "The conduct of this highly technical war produced a great experience in teamwork and professional partnership," Bush writes in an address delivered to social scientists two weeks after the Allies declared victory in Europe in May of 1945.

For Bush, the term "partnership" was vital because in his understanding of techno-scientific mobilization, soldiers and technologists should view themselves as equals. And here he highlights the primacy of "technicians" in the process of innovation because, no matter the scientific or engineering elegance of a new product or process, ultimately the value of the innovation would be judged by the contribution to outcomes on the battlefield. For scientists and engineers, meanwhile, equality between them and elected officials and military officers was vital to protecting and sustaining intellectual independence; however, maintaining a distinction between scientists and engineers, in the heat of war, was impossible. "Many scientists became engineers and some engineers became scientists in the process" of achieving desired results, Bush insists. Both communities also proved able partners with soldiers, he explains.

To this day, Bush's partnership ideal remains the basis for collaboration between civilians and the military.

TEAMWORK OF TECHNICIANS

When I first read of surrendered U-boats proceeding into Allied ports, my mind turned back to 1942 when it was nip and tuck, and the question of whether we could maintain our contact with the United Kingdom and gradually build up there our forces of attack, revolved about the outcome of a technical race between the U-boat men and our means of combatting it. The outcome depended very largely on whether there should arrive first on the scene an increasing U-boat fleet with new devices for eluding attack, remaining submerged for long periods, and with new means of attacking the escorts of convoys; or an enlarged fleet of anti-submarine carriers, long-range aircraft, and surface craft for combating them, with means of finding U-boats in the broad ranges of the Atlantic, pursuing them relentlessly whether they remained on the surface or submerged, and attacking and destroying them by powerful methods that could fully overcome their potent ability to fight back. The outcome was quite conclusive and can be simply summarized. The anti-U-boat means and devices arrived first, and the sinkings went down rapidly during the spring of 1943, and were held down by persistence and vigilance to the end. The expected enemy devices also appeared, but too late. The courage, endurance and sacrifice of the British and Americans on the sea and in the air, armed with the best that science and technology could provide, won a great campaign, one of several, all of which had to be won if our civilization was to survive.

The other side of the record was impressed on me when I read the accounts of the furious struggles on Okinawa, Iwo Jima, and other Japanese-held

Bush gave this address to the National Institute of Social Sciences, May 23, 1945.

islands in the Pacific. By reason of far better implements of war, far better skill in their use, and masterly strategy and tactics, we successfully assaulted the strongest outposts of the Japanese empire. In spite of most formidable fortifications and Nipponese fanaticism expressed especially in the use of suicide bombers operating from their home bases, we progressed on such a basis that our total casualties—on land, at sea and in the air—were far less than the number of the enemy that we had to kill in overcoming their desperate resistance. Our losses were grievous but the magnitude of the accomplishment is hardly yet realized by the people of this country.

The struggles did not end on the battlefield. Tens of thousands of American boys were wounded and the effort continued far into the rear areas and the bases, to save life and restore function. The record of the second aspect of the application of science to warfare is magnificent. Medical aid, with the most advanced methods, was available at the moment a wounded man dropped. Of the seriously wounded that arrived at base hospitals only a few indeed finally succumbed. These base hospitals were often located in tropical areas that were once disease-ridden and had been largely freed from this scourge. Penicillin, blood plasma, sulphonamides [synthetic medicines], DDT, atabrine [an anti-malarial compound], new treatments for burns, advanced surgery, combined with superb organization and rare devotion to duty to bring to bear every aid that science could offer to heal those who carried the fight directly to the enemy, for us all.

But I cannot recount in detail all of these things. They should be told in due time by those who have directly participated in their accomplishment. I can, however, point out one or two threads that ran through the entire effort, and that should never be lost sight of.

The conduct of this highly technical war produced a great experience in teamwork and professional partnership. Under the stress of war, and in the common interest, men always forget their minor differences and sink their personal ambitions and inclinations in the heat and intensity of a hazardous effort. But this has never occurred before, anywhere or during history, to the extent that it occurred this time in the manner in which the United States conducted its enormous war effort all over the world, and it will pay us well to inquire why and how this occurred and what this may augur for the future.

I cannot provide a full analysis here; it will take the lapse of time and the perspective of detachment fully to weigh the factors that have influenced these trends. Moreover we, who have been close to the center, where military and civilian needs become balanced, where over-complex organization creaks as it tries to adapt, see far more of our share of disagreements, petty ambitions and selfish grasping which are on the fringes. The press still finds a story when two men disagree violently in public, and no story when they work in harmony. Those in the field or the laboratory can speak of the solid core on which these minor flaws appear. Their testimony will be that, in spite of all the frailties of human nature, in spite of inevitable cross purposes, there has been a demonstration of collaboration on a scale and to a degree such as has never been seen before. It has been particularly striking in the scientific field.

Military men and scientists arrived at a basis of professional partnership which was extraordinary. Scientists and engineers pulled in the same harness; in fact, many scientists became engineers and some engineers became scientists in the process. Scientists from universities and those from commercial laboratories were indistinguishable. Industrialists, government representatives, officers of the [Armed] Services argued volubly, as red-blooded men should, but they pulled in the same direction in the end and admirable accomplishments were made.

Moreover, the relations between American scientists and those of the British Commonwealth of Nations were just as effective as those within this country, and this is important. Neither group was perfect, of course, but there was just about as much lack of real friction in joint undertakings as in those of a single nationality.

The scientists of this country await a time when they can pay full tribute to the spirit and accomplishments of the scientists of the United Kingdom. They started earlier than we and had the longer stress. They often worked under limitations due to shortages of material, and they were bombed. In the early days of the war, their vision, and the wise utilization of their efforts on radar and in fighter aircraft, enabled the few to turn back the utmost effort of the Luftwaffe to save Britain, and hence ourselves, and to earn the eternal gratitude of the many. In later years, American and British scientists worked so closely together that it will be utterly impossible, and a matter of no vital interest, to attempt to assign many explicit accomplishments to one or the other. In this country we greatly admire British scientists, not only for their eminence in their fields, but for their human qualities. We are proud to have collaborated with them in our joint effort, and we trust that the ties that have been formed will never lessen in strength as we turn toward the advancement of science for the good of humanity in a peaceful world.

I was with Churchill when news from Italy came in, and when General Eisenhower pointed out in his dispatches the great obstacles that had confronted British troops. . . . The Prime Minister quoted these dispatches and said, "it warms the cockles of my heart." So too in viewing the interchange and the community of scientific effort on the two sides of the Atlantic, our hearts in this country are warmed, and we look to the future with confidence that the underlying spirit will endure.

But back of this phenomenon, back of all the teamwork, must lie some great and important causes. I think I have found at least part of the answer, why the two great democracies have given such a heartening exemplification of teamwork and have thus produced great results.

Recently I have been reviewing the progress of the German scientific and technical effort during the war. The account is still fragmentary, and a great part of it remains confidential but some patterns and relations begin to emerge. I have already come to very definite conclusions in my own mind. They are striking and I believe they will be fully borne out when the full history is written.

If a modern scientific war must be fought, the most effective way in which to fight it is under the temporary rigid controls which a continuing democracy voluntarily imposes upon itself as it girds for combat. Such a regime, all other things being equal, can outclass any despotism in bringing to bear on the struggle the combined efforts of science, industry and military might.

A democracy is efficient in emergency for the spirit which it engenders in its normal course is an essential ingredient of great accomplishment under stress. The old contention that only totalitarianism can cope with the complexities of modern life is a fallacy.

De Tocqueville's assertion that the democratic state is less effective than a despotism in a short war, and more effective in a long war, needs to be supplemented under modern conditions of scientific combat. Total complex warfare so emphasizes the advantages of the voluntary collaboration of free men that the democracy will excel in any war, long or short, unless indeed it is so short sighted as to be caught utterly unprepared.

I present these assertions with the conviction that they will ultimately become fully documented and accepted and with the belief that the full appreciation of them is of the first importance in the years ahead. I might go on and state that democracy is the most efficient form even in days of peace with no emergency in the offing, for I believe it, and I hope this too can be demonstrated. But this last point may be academic for we shall live under the shadow of possible emergency of some sort or other for many years to come.

Germany started her war effort many years before the world became properly alerted to the threat, and it began the development of such things as V-bombs many years before we were similarly at work. In certain areas in which great effort was directed, or in which the Germans had natural aptitude, the technical progress of Germany was great and at times even led us in techniques. But these were isolated and relatively unimportant instances. In substantially every important area of the scientific and technical war effort, the enemy [during the course of World War II] was outclassed by the great democracies.

The reason? There are many reasons of course. In the later part of the war the enemy's effort was disrupted by strategic bombing, fortunately for our interests. Slave labor is not efficient labor. But another reason appears to be fundamental. Germany never established partnership, or anything remotely approaching it, between her military men and her scientists. It never brought its scientists, engineers and industrialists into the common effort with genuine teamwork. Its decisions on scientific programs were made in the pattern of all despotic decisions, without the free play and give and take of independent minds, guided by the scientific truth and not by personal fears or ambitions. It badly fumbled the effort at every point. Considering the basis on which that effort was built, one could hardly expect that it would do otherwise.

The German failure was due to many causes. Its comprehensive failure in the scientific field was due in no small degree to the fact that true scientific progress, and its effective utilization, prosper well only in the atmosphere of untrammeled scientific freedom. This is only a small part of the great truth that man reaches his peak accomplishment of mind and intellect only when free, but it is an important point with many practical implications.

With all its enormous advantage of sudden attack upon its peace-loving neighbors, with all its totalitarian regimentation of an entire people for a decade to preparation for the assault, Germany failed. On the scientific front it produced some spectacular and deadly weapons, but it failed in the long arduous race with the great democracies as these applied their scientific accumulations and abilities to their defense. It failed because the atmosphere of freedom is favorable to that collaboration of men of diverse talents which is essential to the effective prosecution of highly complex undertakings. It failed because democracy is the more efficient form; because, when it girds itself for war, it combines the rigid controls which are then essential with the spirit of freedom which it carries over into the emergency. It failed because the regularity and lack of confusion which are the pride of the totalitarians are far exceeded in importance in the modern complex world by the effectiveness, by the efficiency, of the untrammeled spirit which develops fully under freedom.

21

AS WE MAY THINK (1945)

Bush's most significant essay on the future of computing, and the nature of information and its relationship to individual cognition and to human civilization at scale, was published in July 1945 in one of America's most important literary magazines, the Atlantic Monthly. *A richly illustrated version of "As We May Think" appeared in* Life, *America's most popular weekly magazine, later the same year. The essay flows out of a growing awareness of the immense challenges presented by an explosion of information produced by the prosecution of a world war and by the powerful trend toward specialization, which Bush concedes is "increasingly necessary for progress." Bush presents an individual and a collective response to this information explosion, which together, in the words of information historian Robin Boast, amount to both an "ambitious call for a new technology of memory" and a radical leap forward in aspirations for how computers of tomorrow will enhance human problem-solving by imitating the human brain, particularly through "trails" created by cognitive connections, or associations between pieces of information, images, facts, and insights. As envisioned by Bush, the trails of a specific thinker will be stored on a desktop computer and shareable with other people, potentially accelerating the search for new solutions to complex problems.*

Unlike his writing on contemporary political and military affairs, Bush's ideas in "As We May Think" look far into the future of human consciousness and examine novel ways to enhance the value of information and analysis, at an individual and collective level. The essay proved popular with a public captivated by Bush's "endless frontier" concept of a human future animated by technological discovery and exploration. And for the rest of twentieth century, Bush's vision inspired and influenced designers of computers and software, especially in search and retrieval of text and images, today a pervasive aspect of human experience. From customers of Amazon to curious searchers on Google, Bush's principle of organizing information through the mind's manner of association is a cornerstone of the computational world. As he writes in section 6, "Memex Instead of Index," the human mind "operates by association. With one item in its grasp," the mind "snaps instantly to the next [item] that is suggested by the association of thoughts, in accordance with some intricate web of trails carried by the cells of the brain." The problem is that "memory is transitory," and "trails that are not frequently followed are prone to fade," but Bush has a technological fix in the form of a novel machine he calls the memex (short for memory extender).

Leading computer and software innovators, as well as some cognitive and neuroscientists, view Bush's memex idea as anticipating and shaping the development of personal computers and computer networks in the 1970s and beyond. In 1996, the New York Times *counted this essay as among four classic pieces of writing that defined the conceptual contours of computing (the other essays were written by Charles Babbage, Isaac Asimov, and Alan Turing). The* Times *credited Bush with "numerous ideas that have now become common: hypertext, windows, web browsers, document scanning and voice recognition. His vision is one of an almost organic connection between humans and information machines."*

Even in the twenty-first century, Bush's insistence on the intimate relationship between tools of thought and the productivity of the individual mind fuels the ambitions of digital innovators. Many credit Bush with inspiring them to build ever-more powerful and precise information appliances to amplify and extend the power of human cognition through the assistance of personal digital devices connected to powerful data networks (whether located in the cloud or on the World Wide Web). Some even consider Apple's iPhone and other smart phones as the ultimate embodiment of Bush's vision.

The editor of the Atlantic, *Edward Weeks, who led the magazine from 1938 to 1966, penned this brief introduction to "As We May Think," in which he compared Bush to the nineteenth century literary giant, Ralph Waldo Emerson. Weeks's words still stand as a fitting introduction to this seminal essay:*

"Director of the Office of Scientific Research and Development, Dr. Vannevar Bush has coordinated the activities of some six thousand leading American scientists in the application of science to warfare. In this significant article, he holds up an incentive for scientists when the fighting has ceased. He urges that men of science should then turn to the massive task of making more accessible our bewildering store of knowledge. For years inventions have extended man's physical powers rather than the powers of his mind. Trip hammers that multiply the fists, microscopes that sharpen the eye, and engines of destruction and detection are new results, but not the end results, of modern science. Now, says Dr. Bush, instruments are at hand which, if properly developed, will give man access to and command over the inherited knowledge of the ages. The perfection of these pacific instruments should be the first objective of our scientists as they emerge from their war work. Like Emerson's famous address of 1837 on 'The American Scholar,' this paper by Dr. Bush calls for a new relationship between thinking man and the sum of our knowledge."

AS WE MAY THINK

World War II was not a scientist's war; it was a war in which all had a part. The scientists, burying their old professional competition in the demand of a common cause, have shared greatly and learned much. It was

---

"As We May Think" appeared in the July 1945 issue of the *Atlantic Monthly*, reprinted in *Life* magazine, September 10, 1945, and a revised version appeared in *Endless Horizons*, a collection of Bush's essays published in 1946 (Public Affairs Press). I have chosen to rely on the revised 1946 version for two reasons. First, to my eyes, the 1946 version contains changes from the *Atlantic* original that distinctly enhance clarity and meaning. Second, according to the acknowledgements in *Endless Horizon*, Bush "adapted" the text in response to "those considerations which are of equal importance at the present time." Most significantly, the changes allowed Bush to include catchy phrases to open each section, and to account for the end of World War II and the resumption of civilian life in peacetime. By 1946, with the world no longer at war, the pace of civilian technological advance resumed, and Bush grew even more confident in the inevitability of the automation of handling information.

exhilarating to work in effective partnership. Now, for many, this appears to be approaching an end. What are the scientists to do next?

For the biologists, and particularly for the medical scientists, there can be little indecision, for their war has hardly required them to leave the old paths. Many indeed were able to carry on their war research in their familiar peacetime laboratories. Their objectives remain much the same.

It is the physicists who have been thrown most violently off stride, who have left academic pursuits for the making of strange destructive gadgets, who have had to devise new methods for their unanticipated assignments. They did their part on the devices that made it possible to turn back the enemy. They worked in combined effort with the physicists of our allies. They felt within themselves the stir of achievement. They have been part of a great team. Now, as peace approaches, one asks where they will find objectives worthy of their best.

THE BENEFITS OF SCIENCE

Of what lasting benefit has been man's use of science and of the new instruments which his research brought into existence? First, they have increased his control of his material environment. They have improved his food, his clothing, his shelter; they have increased his security and released him partly from the bondage of bare existence. They have given him increased knowledge of his own biological processes so that he has had a progressive freedom from disease and an increased span of life. They are illuminating the interactions of his physiological and psychological functions, giving the promise of an improved mental health.

Science has provided the swiftest communication between individuals; it has provided a record of ideas and has enabled man to manipulate and to make extracts from that record so that knowledge evolves and endures throughout the life of a race rather than that of an individual.

There is a growing mountain of research. But there is increased evidence that we are being bogged down today as specialization extends. The investigator is staggered by the findings and conclusions of thousands of other workers—many of which he cannot find time to grasp, much less to remember, as they appear. Yet specialization becomes increasingly necessary for progress, and the effort to bridge between disciplines is correspondingly superficial.

Professionally our methods of transmitting and reviewing the results of research are generations old and by now are totally inadequate for their purpose. If the aggregate time spent in writing scholarly works and in reading them could be evaluated, the ratio between these amounts of time might well be startling. Those who conscientiously attempt to keep abreast of current thought, even in restricted fields, by close and continuous reading might well shy away from an examination calculated to show how much of the previous month's efforts could be produced on call. Mendel's concept of the laws of genetics was lost to the world for a generation because his publication did not reach the few who were capable of grasping and extending it; and this sort of catastrophe is undoubtedly being repeated all about us, as truly significant attainments become lost in the mass of the inconsequential.

The difficulty seems to be, not so much that we publish unduly in view of the extent and variety of present-day interests, but rather that publication has been extended far beyond our present ability to make real use of the record. The summation of human experience is being expanded at a prodigious rate, and the means we use for threading through the consequent maze to the momentarily important item is the same as was used in the days of square-rigged ships.

But there are signs of a change as new and powerful instrumentalities come into use. Photocells capable of seeing things in a physical sense, advanced photography which can record what is seen or even what is not, thermionic tubes capable of controlling potent forces under the guidance of less power than a mosquito uses to vibrate his wings, cathode ray tubes rendering visible an occurrence so brief that by comparison a microsecond is a long time, relay combinations which will carry out involved sequences of movements more reliably than any human operator and thousands of times as fast—there are plenty of mechanical aids with which to effect a transformation in scientific records.

Two centuries ago Leibnitz invented a calculating machine which embodied most of the essential features of recent keyboard devices, but it could not then come into use. The economics of the situation were against it: the labor involved in constructing it, before the days of mass production, exceeded the labor to be saved by its use, since all it could accomplish could be duplicated by sufficient use of pencil and paper. Moreover, it would have been subject to frequent breakdown, so that it could not have been depended upon; for at that time and long after, complexity and unreliability were synonymous.

Babbage, even with remarkably generous support for his time, could not produce his great arithmetical machine. His idea was sound enough, but construction and maintenance costs were then too heavy. Had a Pharaoh been given detailed and explicit designs of an automobile, and had he understood them completely, it would have taxed the resources of his kingdom to have fashioned the thousands of parts for a single car, and that car would have broken down on the first trip to Giza.

Machines with interchangeable parts can now be constructed with great economy of effort. In spite of much complexity, they perform reliably. Witness the humble typewriter, or the movie camera, or the automobile. Electrical contacts have ceased to stick when thoroughly understood. Note the automatic telephone exchange, which has hundreds of thousands of such contacts, and yet is reliable. A spider web of metal, sealed in a thin glass container, a wire heated to brilliant glow, in short, the thermionic tube of radio sets, is made by the hundred million, tossed about in packages, plugged into sockets—and it works! Its gossamer parts, the precise location and alignment involved in its construction, would have occupied a master craftsman of the guild for months; now it is built for thirty cents. The world has arrived at an age of cheap complex devices of great reliability; and something is bound to come of it.

RECORDING DEVICES

A record, if it is to be useful to science, must be continuously extended, it must be stored, and above all it must be consulted. Today we make the record conventionally by writing and photography, followed by printing; but we also record on film, on wax disks, and on magnetic wires. Even if utterly new recording procedures do not appear, these present ones are certainly in the process of modification and extension.

Certainly progress in photography is not going to stop. Faster material and lenses, more automatic cameras, finer-grained sensitive compounds to allow an extension of the mini-camera idea, are all imminent. Let us project this trend ahead to a logical, if not inevitable, outcome. The camera hound of the future wears on his forehead a lump a little larger than a walnut. It takes pictures 3 millimeters square, later to be projected or enlarged, which after all involves only a factor of 10 beyond present practice. The lens is of universal focus, down to any distance accommodated by the unaided eye, simply because it is of short focal length. There is a built-in photocell on the walnut such as we now have on at least one camera, which automatically adjusts exposure for a wide range of illumination. There is film in the walnut for a hundred exposures, and the spring for operating its shutter and shifting its film is wound once for all when the film clip is inserted. It produces its result in full color. It may well be stereoscopic, and record with two spaced glass eyes, for striking improvements in stereoscopic technique are just around the corner.

The cord which trips its shutter may reach down a man's sleeve within easy reach of his fingers. A quick squeeze, and the picture is taken. On a pair of ordinary glasses is a square of fine lines near the top of one lens, where it is out of the way of ordinary vision. When an object appears in that square, it is lined up for its picture. As the scientist of the future moves about the laboratory or the field, every time he looks at something worthy of the record, he trips the shutter and in it goes, without even an audible click. Is this all fantastic? The only fantastic thing about it is the idea of making as many pictures as would result from its use.

Will there be dry photography? It is already here in two forms. When [Mathew] Brady made his Civil War pictures, the plate had to be wet at the time of exposure. Now it has to be wet during development instead. In the future perhaps it need not be wetted at all. There have long been films impregnated with diazo dyes which form a picture without development, so that it is already there as soon as the camera has been operated. An exposure to ammonia gas destroys the unexposed dye, and the picture can then be taken out into the light and examined. The process is now slow, but someone may speed it up, and it has no grain difficulties such as now keep photographic researchers busy. Often it would be advantageous to be able to snap the camera and to look at the picture immediately.

Another process now in use is also slow, and more or less clumsy. For fifty years impregnated papers have been used which turn dark at every point where an electrical contact touches them, by reason of the chemical change thus produced in an iodine compound included in the paper. They have been used to make records, for a pointer moving across them can leave a trail behind. If the electrical potential on the pointer is varied as it moves, the line becomes light or dark in accordance with the potential.

This scheme is now used in facsimile transmission. The pointer draws a set of closely spaced lines across the paper one after another. As it moves, its potential is varied in accordance with a varying current received over wires from a distant station, where these variations are produced by a photocell which is similarly scanning a picture. At every instant the darkness of the line being drawn is made equal to the darkness of the point on the picture being observed by the photocell. Thus, when the whole picture has been covered, a replica appears at the receiving end.

A scene itself can be just as well looked over line by line by the photocell in this way as can a photograph of the scene. This whole apparatus constitutes a camera, with the added feature, which can be dispensed with if desired, of making its picture at a distance. It is slow, and the picture is poor in detail. Still, it does give another process of dry photography, in which the picture is finished as soon as it is taken.

It would be a brave man who would predict that such a process will always remain clumsy, slow, and faulty in detail. Television equipment today transmits sixteen reasonably good pictures a second, and it involves only two essential differences from the process described above. For one, the record is made by a moving beam of electrons rather than a moving pointer, for the reason that an electron beam can sweep across the picture very rapidly indeed. The other difference involves merely the use of a screen which glows momentarily when the electrons hit, rather than a chemically treated paper or film which is permanently altered. This speed is necessary in television, for motion pictures rather than stills are the object.

Use chemically treated film in place of the glowing screen, allow the apparatus to transmit one picture only rather than a succession, and a rapid camera for dry photography results. The treated film needs to be far faster in action than present examples, but it probably could be. More serious is the objection that this scheme would involve putting the film inside a vacuum chamber, for electron beams behave normally only in such a rarefied environment. This difficulty could be avoided by allowing the electron beam to play on one side of a partition, and by pressing the film against the other side, if this partition were such as to allow the electrons to go through perpendicular to its surface, and to prevent them from spreading out sideways. Such partitions, in crude form, could certainly be constructed, and they will hardly hold up the general development.

Like dry photography, microphotography still has a long way to go. The basic scheme of reducing the size of the record, and examining it by projection rather than directly, has possibilities too great to be ignored. The combination of optical projection and photographic reduction is already producing some results in microfilm for scholarly purposes, and the potentialities are highly suggestive. Today, with microfilm, reductions by a linear factor of 20 can be employed and still produce full clarity when the material is re-enlarged for examination. The limits are set by the graininess of the film, the excellence of the optical system, and the efficiency of the light sources employed. All of these are rapidly improving.

Assume a linear ratio of 100 for future use. Consider film of the same thickness as paper, although thinner film will certainly be usable. Even under these conditions there would be a total factor of 10,000 between the bulk of the ordinary record on books, and its microfilm replica. The *Encyclopedia Britannica* could be reduced to the volume of a matchbox. A library of a million volumes could be compressed into one end of a desk. If the human race has produced since the invention of movable type a total record, in the form of magazines, newspapers, books, tracts, advertising blurbs, correspondence, having a volume corresponding to a billion books, the whole affair, assembled and compressed, could be lugged off in a moving van. Mere compression, of course, is not enough; one needs not only to make and store a record but also be able to consult it, and this aspect of the matter comes later. Even the modern great library is not generally consulted; it is nibbled at by a few.

Compression is important, however, when it comes to costs. The material for the microfilm *Britannica* would cost a nickel, and it could be mailed anywhere for a cent. What would it cost to print a million copies? To print a sheet of newspaper, in a large edition, costs a small fraction of a cent. The entire material of the *Britannica* in reduced microfilm form would go on a sheet eight and one-half by eleven inches. Once it is available, with the photographic reproduction methods of the future, duplicates in large quantities could probably be turned out for a cent apiece beyond the cost of materials. The preparation of the original copy? That introduces the next aspect of the subject.

RECORDING PROCESSES

To make the record, we now push a pencil or tap a typewriter. Then comes the process of digestion and correction, followed by an intricate process of typesetting, printing, and distribution. To consider the first stage of the procedure, will the author of the future cease writing by hand or typewriter and talk directly to the record? He does so indirectly, by talking to a stenographer or a wax cylinder; but the elements are all present if he wishes to have his talk directly produce a typed record. All he needs to do is to take advantage of existing mechanisms and to alter his language.

At a recent World Fair, a machine called a Voder was shown. A girl stroked its keys and it emitted recognizable speech. No human vocal chords entered into the procedure at any point; the keys simply combined some electrically produced vibrations and passed these on to a loudspeaker. In the Bell Laboratories there is the converse of this machine, called a Vocoder. The loudspeaker is replaced by a microphone, which picks up sound. Speak to it, and the corresponding keys move. This may be one element of the postulated system.

The other element is found in the stenotype, that somewhat disconcerting device encountered usually at public meetings. A girl strokes its keys languidly and looks about the room and sometimes at the speaker with a disquieting gaze. From it emerges a typed strip which records in a phonetically simplified language a record of what the speaker is supposed to have said. Later this strip is retyped into ordinary language, for in its nascent form it is intelligible only to the initiated. Combine these two elements, let the Vocoder run the stenotype, and the result is a machine which types when talked to.

Our present languages are not especially adapted to this sort of mechanization, it is true. It is strange that the inventors of universal languages have not seized upon the idea of producing one which better fitted the technique for transmitting and recording speech. Mechanization may yet force the issue, especially in the scientific field; whereupon scientific jargon would become still less intelligible to the layman.

One can now picture a future investigator in his laboratory. His hands are free, and he is not anchored. As he moves about and observes, he photographs and comments. Time is automatically recorded to tie the two records together. If he goes into the field, he may be connected by radio to his recorder. As he ponders over his notes in the evening, he again talks his comments into the record. His typed record, as well as his photographs, may both be in miniature, so that he projects them for examination.

Much needs to occur, however, between the collection of data and observations, the extraction of parallel material from the existing record, and the final insertion of new material into the general body of the common record. For mature thought there is no mechanical substitute. But creative thought and essentially repetitive thought are very different things. For the latter there are, and may be, powerful mechanical aids.

Adding a column of figures is a repetitive thought process, and it was long ago properly relegated to the machine. True, the machine is sometimes controlled by a keyboard, and thought of a sort enters in reading the figures and poking the corresponding keys, but even this is avoidable. Machines which will read typed figures by photocells and then depress the corresponding keys may come from combinations of photocells for scanning the type, electric circuits for sorting the consequent variations, and relay circuits for interpreting the result into the action of solenoids to pull the keys down.

All this complication is needed because of the clumsy way in which we have learned to write figures. If we recorded them positionally, simply by the configuration of a set of dots on a card, the automatic reading mechanism would become comparatively simple. In fact, if the dots are holes, we have the punched-card machine long ago produced by [Herman] Hollerith [an American who invented an electromechanical tabulating machine in early 1900s] for the purposes of the census, and now used throughout business. Some types of complex businesses could hardly operate without these machines.

Adding is only one operation. To perform arithmetical computation involves also subtraction, multiplication, and division, and in addition some method for temporary storage of results, removal from storage for further manipulation, and recording of final results by printing. Machines for these purposes are now of two types: keyboard machines for accounting and the like, manually controlled for the insertion of data, and usually automatically controlled as far as the sequence of operations is concerned; and punched-card machines in which separate operations are usually delegated to a series of machines, and the cards then transferred bodily from one to another. Both forms are very useful; but as far as complex computations are concerned, both are still in embryo.

Rapid electrical counting appeared soon after the physicists found it desirable to count cosmic rays. For their own purposes the physicists promptly constructed thermionic-tube equipment capable of counting electrical impulses at the rate of 100,000 a second. The advanced arithmetical machines of the future will be electrical in nature, and they will perform at 100 times present speeds, or more.

Moreover, they will be far more versatile than present commercial machines, so that they may readily be adapted for a wide variety of operations. They will be controlled by a control card or film, they will select their own data and manipulate it in accordance with the instructions thus inserted, they will perform complex arithmetical computations at exceedingly high speeds, and they will record results in such form as to be readily available for distribution or for later further manipulation. Such machines will have enormous appetites. One of them will take instructions and data from a whole roomful of girls armed with simple keyboard punches, and will deliver sheets of computed results every few minutes. There will always be plenty of things to compute in the detailed affairs of millions of people doing complicated things.

SPECIAL MACHINES

The repetitive processes of thought are not confined, however, to matters of arithmetic and statistics. In fact, every time one combines and records facts in accordance with established logical processes, the creative aspect of thinking is concerned only with the selection of the data and the process to be employed, and the manipulation thereafter is repetitive in nature and hence a fit matter to be relegated to the machine. Not so much has been done along these lines, beyond the bounds of arithmetic, as might be done, primarily because of the economics of the situation. The needs of business and the extensive market obviously waiting, assured the advent of mass-produced arithmetical machines just as soon as production methods were sufficiently advanced.

With machines for advanced analysis no such situation existed; for there was and is no extensive market; the users of advanced methods of manipulating data are a very small part of the population. There are, however, machines for solving differential equations—and functional and integral equations, for that matter. There are many special machines, such as the harmonic synthesizer which predicts the tides. There will be many more, appearing certainly first in the hands of the scientist and in small numbers.

If scientific reasoning were limited to the logical processes of arithmetic, we should not get far in our understanding of the physical world. One might as well attempt to grasp the game of poker entirely by the use of the mathematics of probability. The abacus, with its beads strung on parallel wires, led the Arabs to positional numeration and the concept of zero many centuries before the rest of the world; and it was a useful tool—so useful that it still exists.

It is a far cry from the abacus to the modern keyboard accounting machine. It will be an equal step to the arithmetical machine of the future. But even this new machine will not take the scientist where he needs to go. Relief must be secured from laborious detailed manipulation of higher mathematics as well, if the users of it are to free their brains for something more than repetitive detailed transformations in accordance with established rules. A mathematician is not a man who can readily manipulate figures; often he cannot. He is not even a man who can readily perform the transformations of equations by the use of calculus. He is primarily an individual who is skilled in the use of symbolic logic on a high plane, and especially he is a man of intuitive judgment in the choice of the manipulative processes he employs.

All else he should be able to turn over to his mechanism, just as confidently as he turns over the propelling of his car to the intricate mechanism under the hood. Only then will mathematics be practically effective in bringing the growing knowledge of atomistics to the useful solution of the advanced problems of chemistry, metallurgy, and biology. For this reason there will come more machines to handle advanced mathematics for the scientist. Some of them will be sufficiently bizarre to suit the most fastidious connoisseur of the present artifacts of civilization.

SELECTION TECHNIQUES

The scientist, however, is not the only person who manipulates data and examines the world about him by the use of logical processes, although he sometimes preserves this appearance by adopting into the fold anyone who becomes logical, much in the manner in which a British labor leader is elevated to knighthood. Whenever logical processes of thought are employed—that is, whenever thought for a time runs along an accepted groove—there is an opportunity for the machine. Formal logic used to be a keen instrument in the hands of the teacher in his trying of students' souls. It is readily possible to construct a machine which will manipulate premises in accordance with formal logic, simply by the clever use of relay circuits. Put a set of premises into such a device and turn the crank, and it will readily pass out conclusion after conclusion, all in accordance with logical law, and with no more slips than would be expected of a keyboard adding machine.

Logic can become enormously difficult, and it would undoubtedly be well to produce more assurance in its use. The machines for higher analysis have usually been equation solvers. Ideas are beginning to appear for equation transformers, which will rearrange the relationship expressed by an equation in accordance with strict and rather advanced logic. Progress is inhibited by the exceedingly crude way in which mathematicians express their relationships. They employ a symbolism which grew like Topsy and has little consistency; a strange fact in that most logical field.

A new symbolism, probably positional, must apparently precede the reduction of mathematical transformations to machine processes. Then, on beyond the strict logic of the mathematician, lies the application of logic in everyday affairs. We may someday click off arguments on a machine with the same assurance that we now enter sales on a cash register. But the machine of logic will not look like a cash register, even the streamlined model.

So much for the manipulation of ideas and their insertion into the record. Thus far we seem to be worse off than before—for we can enormously extend the record; yet even in its present bulk we can hardly consult it. This is a much larger matter than merely the extraction of data for the purposes of scientific research; it involves the entire process by which man profits by his inheritance of acquired knowledge. The prime action of use is selection, and here we are halting indeed. There may be millions of fine thoughts, and the account of the experience on which they are based, all encased within stone walls of acceptable architectural form; but if the scholar can get at only one a week by diligent search, his syntheses are not likely to keep up with the current scene.

Selection, in this broad sense, is a stone adze in the hands of a cabinet-maker. Yet, in a narrow sense and in other areas, something has already been done mechanically on selection. The personnel officer of a factory drops a stack of a few thousand employee cards into a selecting machine, sets a code in accordance with an established convention, and produces in a short time a list of all employees who live in Trenton and know Spanish. Even such devices are much too slow when it comes, for example, to matching a set of fingerprints with one of five million on file. Selection devices of this sort will soon be speeded up from their present rate of reviewing data at a few hundred a minute. By the use of photocells and microfilm they will survey items at the rate of a thousand a second, and will print out duplicates of those selected.

This process, however, is simple selection: it proceeds by examining in turn every one of a large set of items, and by picking out those which have certain specified characteristics. There is another form of selection best illustrated by the automatic telephone exchange. You dial a number and the machine selects and connects just one of a million possible stations. It does not run over them all. It pays attention only to a class given by a first digit, then only to a subclass of this given by the second digit, and so on; and thus proceeds rapidly and almost unerringly to the selected station. It requires a few seconds to make the selection, although the process could be speeded up if increased speed were economically warranted. If necessary, it could be made extremely fast by substituting thermionic-tube switching for mechanical switching, so that the full selection could be made in one one-hundredth of a second. No one would wish to spend the money necessary to make this change in the telephone system, but the general idea is applicable elsewhere.

Take the prosaic problem of the great department store. Every time a charge sale is made, there are a number of things to be done. The inventory needs to be revised, the salesman needs to be given credit for the sale, the general accounts need an entry, and, most important, the customer needs to be charged. A central records device has been developed in which much of this work is done conveniently. The salesman places on a stand the customer's identification card, his own card, and the card taken from the article sold—all punched cards. When he pulls a lever, contacts are made through the holes, machinery at a central point makes the necessary computations and entries, and the proper receipt is printed for the salesman to pass to the customer.

But there may be ten thousand charge customers doing business with the store, and before the full operation can be completed someone has to select the right card and insert it at the central office. Now rapid selection can slide just the proper card into position in an instant or two, and return it afterward. Another difficulty occurs, however. Someone must read a total on the card, so that the machine can add its computed item to it. Conceivably the cards might be of the dry photography type I have described. Existing totals could then be read by photocell, and the new total entered by an electron beam.

The cards may be in miniature, so that they occupy little space. They must move quickly. They need not be transferred far, but merely into position so that the photocell and recorder can operate on them. Positional dots can enter the data. At the end of the month a machine can readily be made to read these and to print an ordinary bill. With tube selection, in which no mechanical parts are involved in the switches, little time need be occupied in bringing the correct card into use—a second should suffice for the entire operation. The whole record on the card may be made by magnetic dots on a steel sheet if desired, instead of dots to be observed optically, following the scheme by which Poulsen long ago put speech on a magnetic wire. This method has the advantage of simplicity and ease of erasure. By using photography, however, one can arrange to project the record in enlarged form and at a distance by using the process common in television equipment.

One can consider rapid selection of this form and distant projection for other purposes. To be able to key one sheet of a million before an operator in a second or two, with the possibility of then adding notes thereto, is suggestive in many ways. It might even be of use in libraries, but that is another story. At any rate, there are now some interesting combinations possible. One might, for example, speak to a microphone, in the manner described in connection with the speech-controlled typewriter, and thus make his selections. It would certainly beat the usual file clerk.

MEMEX INSTEAD OF INDEX

The real heart of the matter of selection, however, goes deeper than a lag in the adoption of mechanisms by libraries, or a lack of development of devices for their use. Our ineptitude in getting at the record is largely caused by the artificiality of systems of indexing. When data of any sort are placed in storage, they are filed alphabetically or numerically, and information is found (when it is) by tracing it down from subclass to subclass. It can be in only one place, unless duplicates are used; one has to have rules as to which path will locate it, and the rules are cumbersome. Having found one item, moreover, one has to emerge from the system and re-enter on a new path.

The human mind does not work that way. It operates by association. With one item in its grasp, it snaps instantly to the next that is suggested by the association of thoughts, in accordance with some intricate web of trails carried by the cells of the brain. It has other characteristics, of course; trails that are not frequently followed are prone to fade, items are not fully permanent, memory is transitory. Yet the speed of action, the intricacy of trails, the detail of mental pictures, is awe-inspiring beyond all else in nature.

Man cannot hope fully to duplicate this mental process artificially, but he certainly ought to be able to learn from it. In minor ways he may even improve, for his records have relative permanency. The first idea, however, to be drawn from the analogy concerns selection. Selection by association, rather than indexing, may yet be mechanized. One cannot hope thus to equal the speed and flexibility with which the mind follows an associative trail, but it should be possible to beat the mind decisively in regard to the permanence and clarity of the items resurrected from storage.

Consider a future device for individual use, which is a sort of mechanized private file and library. It needs a name, and, to coin one at random, "memex" will do. A memex is a device in which an individual stores all his books, records, and communications, and which is mechanized so that it may be consulted with exceeding speed and flexibility. It is an enlarged intimate supplement to his memory.

It consists of a desk, and while it can presumably be operated from a distance, it is primarily the piece of furniture at which he works. On the top are slanting translucent screens, on which material can be projected for convenient reading. There is a keyboard, and sets of buttons and levers. Otherwise it looks like an ordinary desk.

In one end is the stored material. The matter of bulk is well taken care of by improved microfilm. Only a small part of the interior of the memex is devoted to storage, the rest to mechanism. Yet if the user inserted 5000 pages of material a day it would take him hundreds of years to fill the repository, so he can be profligate and enter material freely.

Most of the memex contents are purchased on microfilm ready for insertion. Books of all sorts, pictures, current periodicals, newspapers, are thus obtained and dropped into place. Business correspondence takes the same path. And there is provision for direct entry. On the top of the memex is a transparent platen. On this are placed longhand notes, photographs, memoranda, all sorts of things. When one is in place, the depression of a lever causes it to be photographed onto the next blank space in a section of the memex film, dry photography being employed.

There is, of course, provision for consultation of the record by the usual scheme of indexing. If the user wishes to consult a certain book, he taps its code on the keyboard, and the title page of the book promptly appears before him, projected onto one of his viewing positions. Frequently used codes are mnemonic, so that he seldom consults his code book; but when he does, a single tap of a key projects it for his use. Moreover, he has supplemental levers. On deflecting one of these levers to the right he runs through the book before him, each page in turn being projected at a speed which just allows a recognizing glance at each. If he deflects it further to the right, he steps through the book 10 pages at a time; still further at 100 pages at a time. Deflection to the left gives him the same control backwards.

A special button transfers him immediately to the first page of the index. Any given book of his library can thus be called up and consulted with far greater facility than if it were taken from a shelf. As he has several projection positions, he can leave one item in position while he calls up another. He can add marginal notes and comments, taking advantage of one possible type of dry photography, and it could even be arranged so that he can do this by a stylus scheme, such as is now employed in the "telautograph" seen in railroad waiting rooms, just as though he had the physical page before him.

ENDLESS TRAILS

All this is conventional, except for the projection forward of present-day mechanisms and gadgetry. It affords an immediate step, however, to associative indexing, the basic idea of which is a provision whereby any item may be caused at will to select immediately and automatically another. This is the essential feature of the memex. The process of tying two items together is the important thing.

When the user is building a trail, he names it, inserts the name in his code book, and taps it out on his keyboard. Before him are the two items to be joined, projected onto adjacent viewing positions. At the bottom of each there are a number of blank code spaces, and a pointer is set to indicate one of these on each item. The user taps a single key, and the items are permanently joined. In each code space appears the code word. Out of view, but also in the code space, is inserted a set of dots for photocell viewing; and on each item these dots by their positions designate the index number of the other item.

Thereafter, at any time, when one of these items is in view, the other can be instantly recalled merely by tapping a button below the corresponding code space. Moreover, when numerous items have been thus joined together to form a trail, they can be reviewed in turn, rapidly or slowly, by deflecting a lever like that used for turning the pages of a book. It is exactly as though the physical items had been gathered together from widely separated sources and bound together to form a new book. It is more than this, for any item can be joined into numerous trails.

The owner of the memex, let us say, is interested in the origin and properties of the bow and arrow. Specifically he is studying why the short Turkish bow was apparently superior to the English long bow in the skirmishes of the Crusades. He has dozens of possibly pertinent books and articles in his memex. First he runs through an encyclopedia, finds an interesting but sketchy article, leaves it projected. Next, in a history, he finds another pertinent item, and ties the two together. Thus he goes, building a trail of many items. Occasionally he inserts a comment of his own, either linking it into the main trail or joining it by a side trail to a particular item. When it becomes evident that the elastic properties of available materials had a great deal to do with the bow, he branches off on a side trail which takes him through textbooks on elasticity and tables of physical constants. He inserts a page of longhand analysis of his own. Thus he builds a trail of his interest through the maze of materials available to him.

And his trails do not fade. Several years later, his talk with a friend turns to the queer ways in which a people resist innovations, even of vital interest. He has an example, in the fact that the outraged Europeans still failed to adopt the Turkish bow. In fact he has a trail on it. A touch brings up the code book. Tapping a few keys projects the head of the trail. A lever runs through it at will, stopping at interesting items, going off on side excursions. It is an interesting trail, pertinent to the discussion. So he sets a reproducer in action, photographs the whole trail out, and passes it to his friend for insertion in his own memex, there to be linked into the more general trail.

HORIZONS UNLIMITED

Wholly new forms of encyclopedias will appear, ready-made with a mesh of associative trails running through them, ready to be dropped into the memex and there amplified. The lawyer has at his touch the associated opinions and decisions of his whole experience, and of the experience of friends and authorities. The patent attorney has on call the millions of issued patents, with familiar trails to every point of his client's interest. The physician, puzzled by a patient's reactions, strikes the trail established in studying an earlier similar case, and runs rapidly through analogous case histories, with side references to the classics for the pertinent anatomy and histology. The chemist, struggling with the synthesis of an organic compound, has all the chemical literature before him in his laboratory, with trails following the analogies of compounds, and side trails to their physical and chemical behavior.

The historian, with a vast chronological account of a people, parallels it with a skip trail which stops only on the salient items, and can follow at any time contemporary trails which lead him all over civilization at a particular epoch. There is a new profession of trail blazers, those who find delight in the task of establishing useful trails through the enormous mass of the common record. The inheritance from the master becomes, not only his additions to the world's record, but for his disciples the entire scaffolding by which they were erected.

Thus science may implement the ways in which man produces, stores, and consults the record of the race. It might be striking to outline the instrumentalities of the future more spectacularly, rather than to stick closely to methods and elements now known and undergoing rapid development, as has been done here. Technical difficulties of all sorts have been ignored, certainly, but also ignored are means as yet unknown which may come any day to accelerate technical progress as violently as did the advent of the thermionic tube. In order that the picture may not be too commonplace, by reason of sticking to present-day patterns, it may be well to mention one such possibility, not to prophesy but merely to suggest, for prophecy based on extension of the known has substance, while prophecy founded on the unknown is only a doubly involved guess.

All our steps in creating or absorbing material of the record proceed through one of the senses—the tactile when we touch keys, the oral when we speak or listen, the visual when we read. Is it not possible that someday the path may be established more directly?

We know that when the eye sees, all the consequent information is transmitted to the brain by means of electrical vibrations in the channel of the optic nerve. This is an exact analogy with the electrical vibrations which occur in the cable of a television set: they convey the picture from the photocells which see it to the radio transmitter from which it is broadcast. We know further that if we can approach that cable with the proper instruments, we do not need to touch it; we can pick up those vibrations by electrical induction and thus discover and reproduce the scene which is being transmitted, just as a telephone wire may be tapped for its message.

The impulses which flow in the arm nerves of a typist convey to her fingers the translated information which reaches her eye or ear, in order that the fingers may be caused to strike the proper keys. Might not these currents be intercepted, either in the original form in which information is conveyed to the brain, or in the marvelously metamorphosed form in which they then proceed to the hand?

By bone conduction we already introduce sounds into the nerve channels of the deaf in order that they may hear. Is it not possible that we may learn to introduce them without the present cumbersomeness of first transforming electrical vibrations to mechanical ones, which the human mechanism promptly transforms back to the electrical form? With a couple of electrodes on the skull the encephalograph now produces pen-and-ink traces which bear some relation to the electrical phenomena going on in the brain itself. True, the record is unintelligible, except as it points out certain gross misfunctioning of the cerebral mechanism; but who would now place bounds on where such a thing may lead?

In the outside world, all forms of intelligence, whether of sound or sight, have been reduced to the form of varying currents in an electric circuit in order that they may be transmitted. Inside the human frame exactly the same sort of process occurs. Must we always transform to mechanical movements in order to proceed from one electrical phenomenon to another? It is a suggestive thought, but it hardly warrants prediction without losing touch with reality and immediateness.

Presumably man's spirit should be elevated if he can better review his shady past and analyze more completely and objectively his present problems. He has built a civilization so complex that he needs to mechanize his records more fully if he is to push his experiment to its logical conclusion and not merely become bogged down part way there by overtaxing his limited memory. His excursions may be more enjoyable if he can reacquire the privilege of forgetting the manifold things he does not need to have immediately at hand, with some assurance that he can find them again if they prove important.

The applications of science have built man a well-supplied house, and are teaching him to live healthily therein. They have enabled him to throw masses of people against one another with cruel weapons. They may yet allow him truly to encompass the great record and to grow in the wisdom of race experience. He may perish in conflict before he learns to wield that record for his true good. Yet, in the application of science to the needs and desires of [humanity], this would seem to be a singularly unfortunate stage at which to terminate the process, or to lose hope as to the outcome.

22

"LETTER OF TRANSMITTAL" TO PRESIDENT HARRY TRUMAN (1945)

Bush's best-known piece of political writing is his summary to the Endless Frontier *report on the future relations of the U.S. government to science, issued in July 1945. Bush commissioned the report believing that President Franklin Roosevelt would be his primary audience. Bush had grown fond of Roosevelt during the war years. Roosevelt was decisive, clear, and secretive, and Bush favored expressing himself on government policy matters in private. Bush believed that because the decisive role of U.S.-funded science and technology in the defeat of Germany and Japan that academic scientists and engineers should receive generous government support following the war. Roosevelt's death in April 1945 robbed Bush of his strongest ally in government. President Harry Truman, who had served as vice president for only a few months, was an unknown quantity and Bush's language in this letter of July 5, 1945, suggests the need to aim to tutor the new president on the value of research and how science as an active pursuit, carried out at scale, can nourish "the pioneer spirit . . . still vigorous within our nation."*

"LETTER OF TRANSMITTAL," *SCIENCE, THE ENDLESS FRONTIER*

Dear Mr. President:

In a letter dated Nov. 17, 1944, President Roosevelt requested my recommendations on the following points:

(1) What can be done, consistent with military security, and with the prior approval of the military authorities, to make known to the world as soon as possible the contributions which have been made during our effort to scientific knowledge?

(2) With particular reference to the war of science against disease, what can be done now to organize a program for continuing in the future the work which has been done in medicine and related sciences?

(3) What can government do now and in the future to aid research activities by public and private organizations?

(4) Can an effective program be proposed for discovering and developing scientific talent in American youth so that the continuing future of scientific research in this country may be assured on a level comparable to what has been done during the war?

It is clear from President Roosevelt's letter that in thinking of science he had in mind the natural sciences, including biology and medicine, and I have so interpreted his questions. Progress in other fields, such as the social sciences and the humanities, is likewise important; but the program for science in my reports warrants immediate attention.

In seeking answers to President Roosevelt's questions I have had the assistance of distinguished committees specially qualified to advise in respect to these subjects. The committees have given these matters the serious attention they deserve; indeed, they have regarded this as an opportunity to participate in shaping the policy of the country with reference to scientific research. They have had many meetings and have submitted formal reports. I have been in close touch with the work of the committees and with their members throughout. I have examined all of

From *Science, the Endless Frontier: A Report to the President on a Program for Postwar Scientific Research*, by Vannevar Bush, Director of the Office of Scientific Research and Development, July 1945.

the data they assembled and the suggestions they submitted on the points raised in President Roosevelt's letter.

Although the report which I submit herewith is my own, the facts, conclusions, and recommendations are based on the findings of the committees. Since my report is necessarily brief, I am including as appendices the full reports of the committees.

A single mechanism for implementing the recommendations of the several committees is essential. In proposing such a mechanism I have departed somewhat from the specific recommendations of the committees, but I have since been assured that the plan I am proposing is fully acceptable to the committee members.

The pioneer spirit is still vigorous within this nation. Science offers a largely unexplored hinterland for the pioneer who has the tools for his task. The rewards of such exploration, both for the nation and the individual, are great. Scientific progress is one essential key to our security as a nation, to our better health, to more jobs, to a higher standard of living, and to our cultural progress.

Respectfully yours,

V. Bush

23

"SUMMARY" OF *SCIENCE, THE ENDLESS FRONTIER* (1945)

The summary, or introduction, written by Bush as an opening to the report entitled Science, the Endless Frontier, *stands among the important pieces of American political writing in the twentieth century. To this day, Bush's justification for public funding of research remains popular. The connection he makes between the pursuit of knowledge and the exploration of geographic frontiers has long resonated with ordinary Americans, researchers, and politicians alike. Worth noting is Bush's insistence that the summary to* Endless Frontier *"is my own," though the bulk of the report is rather technical, dated, and the creation of an array of committees, whose members Bush appointed. In his summary, Bush jammed into a mere two thousand words an essential charter for the political support of research, and he succinctly presented his most cherished policy ideas on innovation, economic growth, national security, and the role of science, engineering, and knowledge in a good society. The summary was written in his voice, and he held to its central positions for the rest of his life. While considered a paean to unfettered scientific research, in his summary Bush also emphasized the potential economic benefits from research, and he anticipated that taxpayers and members of Congress would not long support scientists and engineers unless their new knowledge and techniques translated ultimately into practical and commercial gain. In this spirit, Bush advocated for lifting "the lid" of secrecy on "a vast amount of information relating to the application of science" to real-world problems. "Much of this can be used by industry," he insisted. While viewed today as a justi-fication for federal funding for scientific and technological research, Bush saw the report as an opportunity, a month before the first atomic bomb on Hiroshima, to highlight that innovation now could determine the fate of nations and define global competition for decades to come. He expressed the view, soon to become conventional wisdom in the new "atomic age," that "in large measure" America's future depended on applying smart science and engineering to vital national challenges.*

"SUMMARY" OF *SCIENCE, THE ENDLESS FRONTIER*

Scientific Progress is Essential: Progress in the war against disease depends upon a flow of new scientific knowledge. New products, new industries, and more jobs require continuous additions to knowledge of the laws of nature, and the application of that knowledge to practical purposes. Similarly, our defense against aggression demands new knowledge so that we can develop new and improved weapons. This essential, new knowledge can be obtained only through basic scientific research.

Science can be effective in the national welfare only as a member of a team, whether the conditions be peace or war. But without scientific progress no amount of achievement in other directions can insure our health, prosperity, and security as a nation in the modern world.

For the War Against Disease: We have taken great strides in the war against disease. The death rate for all diseases in the Army, including overseas forces, has been reduced from 14.1 per thousand in the last war to 0.6 per thousand in this war. In the last 40 years life expectancy has increased from 49 to 65 years, largely as a consequence of the reduction in the death rates of infants and children. But we are far from the goal. The annual deaths from one or two diseases far exceed the total number of American lives lost in battle during this war. A large fraction of these deaths in our civilian population cut short the useful lives of our citizens. Approximately 7,000,000 persons in the United States are mentally ill and their care costs

---

From *Science, the Endless Frontier: A Report to the President on a Program for Postwar Scientific Research*, by Vannevar Bush, Director of the Office of Scientific Research and Development, July 1945.

the public over $175,000,000 a year. Clearly much illness remains for which adequate means of prevention and cure are not yet known.

The responsibility for basic research in medicine and the underlying sciences, so essential to progress in the war against disease, falls primarily upon the medical schools and universities. Yet we find that the traditional sources of support for medical research in the medical schools and universities, largely endowment income, foundation grants, and private donations, are diminishing and there is no immediate prospect of a change in this trend. Meanwhile, the cost of medical research has been rising. If we are to maintain the progress in medicine which has marked the last 25 years, the Government should extend financial support to basic medical research in the medical schools and in universities.

For Our National Security: The bitter and dangerous battle against the U-boat was a battle of scientific techniques—and our margin of success was dangerously small. The new eyes which radar has supplied can sometimes be blinded by new scientific developments. V-2 was countered only by capture of the launching sites.

We cannot again rely on our allies to hold off the enemy while we struggle to catch up. There must be more—and more adequate—military research in peacetime. It is essential that the civilian scientists continue in peacetime some portion of those contributions to national security which they have made so effectively during the war. This can best be done through a civilian-controlled organization with close liaison with the Army and Navy, but with funds direct from Congress, and the clear power to initiate military research which will supplement and strengthen that carried on directly under the control of the Army and Navy.

And for the Public Welfare: One of our hopes is that after the war there will be full employment. To reach that goal the full creative and productive energies of the American people must be released. To create more jobs we must make new and better and cheaper products. We want plenty of new, vigorous enterprises. But new products and processes are not born full-grown. They are founded on new principles and new conceptions which in turn result from basic scientific research. Basic scientific research is scientific capital. Moreover, we cannot any longer depend upon Europe as a major source of this scientific capital. Clearly, more and better scientific research is one essential to the achievement of our goal of full employment.

How do we increase this scientific capital? First, we must have plenty of men and women trained in science, for upon them depends both the creation of new knowledge and its application to practical purposes. Second, we must strengthen the centers of basic research which are principally the colleges, universities, and research institutes. These institutions provide the environment which is most conducive to the creation of new scientific knowledge and least under pressure for immediate, tangible results. With some notable exceptions, most research in industry and Government involves application of existing scientific knowledge to practical problems. It is only the colleges, universities, and a few research institutes that devote most of their research efforts to expanding the frontiers of knowledge.

Expenditures for scientific research by industry and Government increased from $140,000,000 in 1930 to $309,000,000 in 1940. Those for the colleges and universities increased from $20,000,000 to $31,000,000, while those for the research institutes declined from $5,200,000 to $4,500,000 during the same period. If the colleges, universities, and research institutes are to meet the rapidly increasing demands of industry and Government for new scientific knowledge, their basic research should be strengthened by use of public funds.

For science to serve as a powerful factor in our national welfare, applied research both in Government and in industry must be vigorous. To improve the quality of scientific research within the Government, steps should be taken to modify the procedures for recruiting, classifying, and compensating scientific personnel in order to reduce the present handicap of governmental scientific bureaus in competing with industry and the universities for top-grade scientific talent. To provide coordination of the common scientific activities of these governmental agencies as to policies and budgets, a permanent Science Advisory Board should be created to advise the executive and legislative branches of Government on these matters.

The most important ways in which the Government can promote industrial research are to increase the flow of new scientific knowledge through support of basic research, and to aid in the development of scientific talent. In addition, the Government should provide suitable incentives to industry to conduct research, (a) by clarification of present uncertainties in the Internal Revenue Code in regard to the deductibility of research and development expenditures as current charges against net income, and (b) by strengthening the patent system so as to eliminate uncertainties which now bear heavily on small industries and so as to prevent abuses which reflect discredit upon a basically sound system. In addition, ways should be found to cause the benefits of basic research to reach industries which do not now utilize new scientific knowledge.

We Must Renew Our Scientific Talent: The responsibility for the creation of new scientific knowledge—and for most of its application—rests on that small body of men and women who understand the fundamental laws of nature and are skilled in the techniques of scientific research. We shall have rapid or slow advance on any scientific frontier depending on the number of highly qualified and trained scientists exploring it.

The deficit of science and technology students who, but for the war, would have received bachelor's degrees is about 150,000. It is estimated that the deficit of those obtaining advanced degrees in these fields will amount in 1955 to about 17,000—for it takes at least 6 years from college entry to achieve a doctor's degree or its equivalent in science or engineering. The real ceiling on our productivity of new scientific knowledge and its application in the war against disease, and the development of new products and new industries, is the number of trained scientists available.

The training of a scientist is a long and expensive process. Studies clearly show that there are talented individuals in every part of the population, but with few exceptions, those without the means of buying higher education go without it. If ability, and not the circumstance of family fortune, determines who shall receive higher education in science, then we shall be assured of constantly improving quality at every level of scientific activity. The Government should provide a reasonable number of undergraduate scholarships and graduate fellowships in order to develop scientific talent in American youth. The plans should be designed to attract into science only that proportion of youthful talent appropriate to the needs of science in relation to the other needs of the nation for high abilities.

Including Those in Uniform: The most immediate prospect of making up the deficit in scientific personnel is to develop the scientific talent in the generation now in uniform. Even if we should start now to train the current crop of high-school graduates, none would complete graduate studies before 1951. The Armed Services should comb their records for men who, prior to or during the war, have given evidence of talent for science, and make prompt arrangements, consistent with current discharge plans, for ordering those who remain in uniform, as soon as militarily possible, to duty at institutions here and overseas where they can continue their scientific education. Moreover, the Services should see that those who study overseas have the benefit of the latest scientific information resulting from research during the war.

The Lid Must Be Lifted: While most of the war research has involved the application of existing scientific knowledge to the problems of war, rather than basic research, there has been accumulated a vast amount of information relating to the application of science to particular problems. Much of this can be used by industry. It is also needed for teaching in the colleges and universities here and in the Armed Forces Institutes overseas. Some of this information must remain secret, but most of it should be made public as soon as there is ground for belief that the enemy will not be able to turn it against us in this war. To select that portion which should be made public, to coordinate its release, and definitely to encourage its publication, a Board composed of Army, Navy, and civilian scientific members should be promptly established.

A Program for Action: The Government should accept new responsibilities for promoting the flow of new scientific knowledge and the development of scientific talent in our youth. These responsibilities are the proper concern of the Government, for they vitally affect our health, our jobs, and our national security. It is in keeping also with basic United States policy that the Government should foster the opening of new frontiers and this is the modern way to do it. For many years the Government has wisely supported research in the agricultural colleges and the benefits have been great. The time has come when such support should be extended to other fields.

The effective discharge of these new responsibilities will require the full attention of some overall agency devoted to that purpose. There is not now in the permanent Governmental structure receiving its funds from Congress an agency adapted to supplementing the support of basic research in the colleges, universities, and research institutes, both in medicine and the natural sciences, adapted to supporting research on new weapons for both Services, or adapted to administering a program of science scholarships and fellowships.

Therefore I recommend that a new agency for these purposes be established. Such an agency should be composed of persons of broad interest and experience, having an understanding of the peculiarities of scientific research and scientific education. It should have stability of funds so that long-range programs may be undertaken. It should recognize that freedom of inquiry must be preserved and should leave internal control of policy, personnel, and the method and scope of research to the institutions in which it is carried on. It should be fully responsible to the President and through him to the Congress for its program.

Early action on these recommendations is imperative if this nation is to meet the challenge of science in the crucial years ahead. On the wisdom with which we bring science to bear in the war against disease, in the creation of new industries, and in the strengthening of our Armed Forces depends in large measure our future as a nation.

24

SOLDIERS AND SCIENTISTS IN PARTNERSHIP (1946)

Bush cared deeply about military-civil relations, and he thought this central relationship in the American republic was misunderstood. He pondered intensely how soldiers and scientists, warriors and technologists, could work well together (or not). While American intellectuals tend to reflect on the daunting challenges of civilian control of the military, Bush, drawing on his experience in World War II, obsessed over how talented civilians, especially those talented in science and engineering, management and organization, could decisively influence the world of the warrior (or not, to the detriment of the latter). In this essay, reviewing the war experience, Bush assessed the value of partnership between soldiers and scientists. If the political theorist Samuel Huntington worried about civil-military relations from the standpoint of democracy and the role of expert advice in political decision-making, Bush viewed the problem of national security as essentially the challenge of organizing the military around new weapons, new tactics, and future innovations. And to achieve this meant the military must have intimate relations with scientists and engineers. Bush wanted the generals and the admirals to anticipate where combat might go, and where and how new security threats might emerge. For Bush, "soldiers and scientists in partnership" was not an empty slogan, but the source of sound military planning and practices.

SOLDIERS AND SCIENTISTS IN PARTNERSHIP

The history of the Office of Scientific Research and Development [is] the history of a rapid transition, from warfare as it has been waged for thousands of years by the direct clash of hordes of armed men, to a new type of warfare in which science becomes applied to destruction on a wholesale basis. It marks, therefore, a turning point in the broad history of civilization.

It begins in 1940, when this country was still asleep under the delusion of isolation—when only a few realized that a supreme test was inevitable, to determine whether the democratic form of government could survive—when none could see clearly the full revolution in the art of war that impended.

[Any history of OSRD] recites the extraordinarily rapid evolution of weapons, as the accumulated backlog of scientific knowledge became directly applied to radar, amphibious warfare, aerial combat, the proximity fuse and the atomic bomb.

But it tells also of something that is more fundamental even than this diversion of the progress of science into methods of destruction. It shows how men of good will, under stress, can outperform all that dictatorship can bring to bear—as they collaborate effectively, and apply those qualities of character developed only under freedom. It demonstrates that democracy is strong and virile, and that free men can defend their ideals as ably in a highly complex world as when they left the plow in the furrow to grasp the smoothbore. This is the heartening fact which should give us renewed courage and assurance, even as we face a future in which war must be abolished, and in which that end can be reached only by resolution, patience and resourcefulness of a whole people.

Not that the great accomplishment of girding a nation for war was automatic or simple, even in this phase. On one hand were military men, burdened with extreme responsibility that only military men can carry of ordering great numbers of their fellows into strife where many of them

---

Bush wrote this essay to serve as the foreword to the official government account of the contributions of science and engineering during World War II, published as *Scientists Against Time* by James Baxter Phinney (Cambridge, Mass.: MIT Press, 1968). The essay is dated January 21, 1946.

must die, harassed by an unreasonable load of work, determined to get on with the tough and appalling job and get it over. On the other hand were scientists and engineers, realizing from their background that much they saw was obsolete, forced to learn overnight a new and strange way of life and set of human relations, driven to the limit by the keen realization of the scientific competence of the enemy and the consequent desperate nature of the race. And between were industrialists, faced with the production of unprecedented quantities of strange devices, wary of the abrupt changes that would wreck the mass production in which this country excelled. These diverse groups and points of view could not collaborate unless they were forced to do so artificially and ineffectively by arbitrary orders from above, or unless they learned a new partnership. They accomplished the latter, and the accomplishment was greater than the mere creation of new weapons, and made such creation possible on a scale which determined the outcome.

There were no disloyalties, or so few as to be negligible. There were few who quailed at the heavy responsibilities. But in every group there are the small men, those whose selfishness persists, those whose minds are frozen, those whose pride is false. There were blocked programs, and futile efforts. There was also plenty of hearty disagreement, and vigorous argument in conference. But these were all incidents. Out of it evolved, toward the end, an effective professional partnership of scientists, engineers, industrialists, and military men, as was never seen before, which exemplified the spirit of America in action at its strongest and best.

This is the story of the development of weapons of war, but it is also the story of an advance in the whole complex of human relations in a free society, and the latter is of the greater significance.

It is also the story of the advance of medicine under war stress. As man's knowledge of his environment extends, that is as science advances, it is well that it should be applied to ease man's lot and fend off the harshness of nature. This is not the greatest goal of the extension of knowledge, but it is a very great one. It was kept in view even in war, while the great weight of science became applied to destruction; in fact it was accelerated. Surgical techniques, blood substitutes, anti-malarial, penicillin, these and others were advanced at far beyond the peacetime rate. In fact, in the long run, it was probable that the medical advance of the war will save as many lives as were lost by military operations during its continuance. This is the mission of the medical profession, to save lives and mitigate suffering, and it was well done.

The same group of scientists and engineers who banded themselves together in 1940 stayed together and finished the job. Most of them have now returned to the peacetime tasks which were interrupted, to the extension of knowledge and the training of the next generation. They take with them justifiable pride in the history which is here recited, and many deep and abiding friendships, forged under stress, among themselves and with their partners of the Army and Navy. Many have specific accomplishments to cherish, successes for which they took full responsibility; for the entire success of OSRD depended upon extreme decentralization and great autonomy of individual units. All have the satisfaction of having been members of one of the finest teams of men ever assembled in a great cause.

25

ORGANIZING SCIENTIFIC RESEARCH
FOR WAR (1946)

Here is Bush's pithy case for the necessity for the nation-state to closely link war and innovation. How to do so, consistent with democracy and with the imperative of improving the quality of life for citizens, was a core concern for Bush. Could soldiers and scientists work as equals, and how much of the benefits of innovation ought to be sacrificed in the name of defense against enemies real and imagined? To this thorny subject Bush often turned, and in this brief, eloquent summary he anticipates the argument that the rise of a military-industrial complex is both inevitable and impossible without the participation of leading scientists. World War II brought, Bush writes, "closer linkage among military men, scientists, and industrialists" than the nation had ever seen. Fifteen years before President Eisenhower bemoaned the threat to democracy of the military-industrial complex, Bush strongly hinted that such a linkage would become the new normal, "as is natural in an essentially scientific and technological age."

ORGANIZING SCIENTIFIC RESEARCH FOR WAR

World War II was the first war in human history to be affected decisively by weapons unknown at the outbreak of hostilities. This is probably the

---

Bush wrote these reflections as the foreword to another official government accounting of science and the war, *Organizing Scientific Research for War: The Administrative History of the Office of Scientific* most significant military fact of our decade: that upon the current evolution of the instrumentalities of war, the strategy and tactics of warfare must now be conditioned. In World War II this new situation demanded a closer linkage among military men, scientists, and industrialists than had ever before been required, primarily because the new weapons whose evolution determines the course of war are dominantly the products of science, as is natural in an essentially scientific and technological age.

The Office of Scientific Research and Development, one crucial aspect of whose history is ably told in this volume, was the medium through which, in the main, scientists were joined in effective partnership with military men.

Such a partnership was really a new thing in the world and was a partnership between groups which one might at first thought consider inherently compatible. The military group, both because of the extreme demands and extreme responsibilities of the profession of arms and because of a long and honorable tradition, is formally organized to a very high degree. The scientific group, both because of the individualistic approach essential to research and because of the sufficiency of the loosest of organization for all practical purposes in normal times, is much more a gathering of individuals than a group in the professional or structural sense.

For two such entities to develop a pattern of highly effective collaboration—and that such a pattern was developed is clear in the record—demanded much in the beginning from each. New lessons in understanding and evaluation had to be learned both by the military and by the scientific. Old preconceptions had to be overcome and old prepossessions foregone by both. In the earlier part of the period it was inevitable that there should be much expenditure of time and energy in these vital developments, inevitable also that there should be disagreement and even outright friction. Honest and strongly held opposing convictions often had to be reconciled. Obduracy and narrow interests occasionally had to be rooted out. Both parties generally, however, were imbued with the same patriotism and actuated by the same sense of urgent responsibility, which speeded the establishment of sound interdependent effort.

Research and Development (Boston: Little, Brown, 1948). This book, authored by Irvin Stewart (president of West Virginia University and Bush's deputy director at the Office of Scientific Research and Development), emphasized management and administration. While Bush signed and dated the text on November 4, 1946, the book wasn't released until 1948 by Atlantic Monthly Press.

It was at the administrative center where all the widely ramified activities and contacts of the scientific effort could be seen from a somewhat detached point of view that the process of adjustment and comprehension was most sharply sensed. Hence it is that [this] account of the administrative history of the OSRD is of the greatest value for the future and of the greatest interest to those concerned with organization and the patterns of government. It offers the data on which may be based sound appraisal of how sincere and hard-working men of professions ordinarily widely separate can set about and accomplish the development of unity.

It was the function of the administrative office of OSRD to channelize and focus an amazing array of variegated activities, to coordinate them both with the military necessities which they were designed to help to meet and with the requirements of the powerful industrial structure on which their effective application relied. In the contracting system which it developed, in the methods for safeguarding the public interest through sound patent policy which it created, in the means for effective and cordial liaison with co-operating agencies which it effected, and in a dozen other ways, the office brought to being a pattern of administration which aptly met a new and unique need and which stands as a richly suggestive guide for other undertakings.

26

THE DANGER OF DICTATION OF SCIENCE BY LAYMEN (1946)

As World War II receded in time and partisan political differences emerged again in America, Bush reminded his audiences that government should refrain from telling scientists what to do, think, research, and publish. His fear might seem overblown to our present time when scientists dissent freely and at times contradict government policy, but after the war, the rising status of the Soviet Union as a technological power spawned new concepts of the relationship between science and society. In the communist framework, scientists were viewed as instruments of the people, and their own aims should be subordinated to the needs of the nation-state. Such notions, however attractive in theory, in practice often meant that scientists and engineers were forced to shape their research in support of the prevailing political agenda. In the American context, the federal government's emerging role as chief funder of fundamental, or pure, scientific research raised anxieties about a loss of independence by scientists and engineers. While Bush worried about government interference, and the perils of political coercion of researchers, he also worried that large corporations, and perhaps wealthy benefactors, might also unduly influence what scientists studied, published, and promoted. In expressing these concerns, Bush was himself influenced by wider fears among intellectuals that "group think" and the "commissar" put at risk independent inquiry, whether techno-scientific or humanistic. Here he especially worries that political control of funds may occasion "injurious dictation to science by laymen."

THE DANGER OF DICTATION OF SCIENCE BY LAYMEN

The war has placed science in an utterly new position. Just what this position is . . . [is] well worth analysis now that we are turning to programs of peace.

The reason there is an altered position for science is very clear. It is now recognized all over the world that the application of science is central in national security. During the second world war, the nature of warfare underwent a complete transformation. This was of much greater scope than could have been indicated by the actual hostilities from 1939 to 1945, for changes in the methods of making war do not produce their full effect at once. It is not sure that great armies and great fleets are obsolete, but it is sure that if there were another great war it would be vastly different in its nature from anything that has preceded. The revolution is greater than that brought about by the advent of gun powder, or that occasioned by the use of shells and the consequent entry of the armored ship. It is in fact greater than the mere influence of a technique for [the revolution] involves rather a method for applying techniques. The results then in the long run can be as far reaching as those which came to pass when men learned that inventions can be made by deliberate planned effort, or when tools were caused to make more tools and thus to multiply themselves. Full comprehension of the fundamental change in all of mankind's concerns which is foretold in the scientific revolution in warfare cannot be had for some time; the implications are too numerous, too involved, too subtle. Such comprehension will come, nevertheless; it is inevitable in view of the way in which the past war was conducted by all participants and the lessons that were thus recorded for all to read.

This present sweeping transformation of mankind's affairs had its beginnings when the first simian developed curiosity, when he came to manipulate the objects in nature about him not for their immediate animal ends but simply because he had a new urge. The appearance of that urge marked the advent of the inchoate human intelligence as a wholly new factor in a hitherto drab evolution. It thus began the great experiment which surely will go on to full and richer scope, for man's urge to

---

Excerpted from Bush's *Report of the President*, Carnegie Institution of Washington Year Book, December 1946.

learn and to know more about his environment and his fellows will not end now. Yet by its intermediate results our days are now perplexed.

The world which faces this multiple problem and opportunity is still in a highly nationalistic phase of organization. It was this fact that Churchill undoubtedly had in mind when he declared the world not yet ready for such an advance as the development of atomic energy. But the world has never been ready in that sense: it was not ready for the advent of the use of fire, or of mass communication. This time, however, there is a difference for there is a general recognition that the advance, though its acceleration of the evolution of weapons, bids fair to proceed so far as to render some form of world unity inevitable. This generally held conviction, however, leads to far different concepts with different groups and different individuals. To many, taking the world as a whole, it points inevitably to a Roman peace, one world under some sort of domination, held together by the very power of the new instrumentalities. But there is a large and salutary endeavor for . . . a voluntary joining of states under some scheme of guarantees that will protect minorities. This is not idle dreaming; the formation of the United States by the voluntarily joining of the colonies, and other unions as well, had their origin in similar efforts. We need not be dismayed even by the raucous accusations and turmoil of the present day, for the colonies in North America before they became states, and even after, indulged in similar recriminations as they worked toward union. Still, anything approaching a unitary world has never yet been achieved on the basis of voluntary association, and the accomplishment will take time. The principal question is whether there will be time enough— whether, in short, there must be another demonstration of the power of applied science to destroy—before mankind as a whole recognizes that a new approach toward international functioning is demanded. This is a serious question. But it is doubtful whether civilization will commit suicide knowing it is doing so, and because of science the race between the power of weapons and the power of understanding is not altogether one-sided.

For the application of science has done more than produce a power of destruction which has made a world organization essential. It has in fact also provided means of primary importance to the development of such organization, in the form of universal communication, the printed word, transmitted speech, swift transport. Real unity can be achieved only through general mass enlightenment throughout the world, and hundreds of millions of people cannot know one another except as new instrumentalities free them from the narrow limits of mere personal contact. It should also be remembered that the application of science can alleviate as well as destroy. In fact the progress in medicine under the forced draft of war in the past conflict was such that the number of lives saved goes far toward balancing battle casualties. Still more important, perhaps, great masses of people now have the definite promise before them of leading healthy lives, especially in the tropics, and a healthy outlook on life is necessary for great movements of collaboration.

All this furnishes the background for the position of science in our own national scene. . . . There is no question that scientific effort in this country will now expand to a far greater proportion of the total volume of effort than was the case before the war. Nor is there a question that as it does so it will receive public attention and support. On the expanding scale which the national interest requires and dictates, federal support of scientific effort now becomes essential and is also inevitable. The form that it may take is of great importance. Many dangers may be discerned here, and it will be well to review some of them.

First there is the danger that there will be over-emphasis on the applied phase for science, for the public is alert to the tangible benefits to be had from it, but hardly realizes the fact that they are all dependent upon long-term advance in fundamental science. Fortunately, in the immediate postwar period at least, this danger has been averted. Federal support to university research is already flowing in large measure, and it is now directed heavily toward support of basic science. But this may be a passing phase, and the danger certainly remains, for as a people we are strongly [technical], we have always excelled in the applied, we have not turned with the same success to more philosophical matters. In many branches of science we have as a nation hence lagged behind Europe.

There is also a danger that control of funds may occasion injurious dictation to science by laymen. The fact that this is a somewhat subtle matter renders the danger much greater. In applying science it is often correct that a group of laymen should set the general objectives—and in industrial research, for example, where men of diverse backgrounds and interest need to meet with the scientists and engineers in order to create a program that is sound from the standpoint of industry. The governing

boards of universities rightly participate as the scientific research programs of their institutions are formulated, not merely because they
must assure that plans and means shall be commensurate but also—and
more importantly—because they must counsel in the defining of broad
objectives which reach beyond the direct interests of the scientists into
the greater question of the ultimate best interests of the country. In both
cases, however, once the general objectives are defined, wise management
leaves the methods of approach entirely to the scientists. The danger is
that this lay participation will go beyond its appropriate function, enter
into the methods themselves, and seek to influence the choice of the particular paths to be followed. If a scientist is really competent in his field, he
knows better than anyone else how, in the exceedingly complex situation
surrounding the frontier of knowledge, to single out an approach which
may lead toward greater attainment. Interference with him by any individual, board or committee, as he thus determines his way, annoys him
greatly and should. The finding of the path is one of the finer parts of his
art; in fact his rise to eminence depends very decidedly upon the wisdom
with which he can thus choose.

   To illustrate, there is today in this country a great urge to clear up once
and for all at least the worst aspects of the great curse of cancer. Moreover
because of recent advances, new approaches of promise exist. Certainly
funds poured into this field at the present time are well invested. Yet how
does one proceed from here? One method favorably known to Americans
because of the great advances which it has produced in applied science is
to assemble a group of highly intelligent citizens, to build great laboratories and install therein competent scientists, and to create patterns of effort
paralleling those that have been successful in large industrial laboratories,
with the single aim of finding a cure. But there is an alternative method,
recommended by its admirable results in fundamental research. This is to
select scientific men of great power—men who are thus regarded by their
colleagues—and see to it that they get every bit of support which they can
utilize effectively, in their own undertakings, and in accordance with their
own plans. Such an effort should cover every contributory field, and hence
the entire science of man's physical and chemical constitution and growth.
It might be that the first method would find a solution—such things do
happen. The question is essentially one of timing. If the investigation of
cancer has come to the stage of applied research, the organized approach

is entirely sound. If that investigation is still in the phase demanding fundamental research—and the evidence emphatically indicates that it is—then the second method is the one to follow. Through it, by and large, have come the great accomplishments in fundamental science, and it is sure to bring results in the long run, in many fields of application at once and over a broad range. The characteristic and productive urge of Americans to move swiftly into applied research for immediate and practical results could easily lead to ignoring the vital fact.

The question of just where governing boards should stop in their proper control of funds as they deal with scientists of eminence has not, however, been a serious problem, for in this country sound understanding on this matter has rather generally been established. It becomes important at the present time because new governmental boards of one sort or another will be controlling federal funds for scientific research, and it is essential that the productive pattern which has been developed in the past should not now be departed from simply because pressure exists [to deliver positive outcomes] and funds are large.

This raises the question of course whether the flow of federal funds into university research will lead to federal domination of universities in one way or another. It need not do so. The history of state universities in this country shows how public funds can be devoted in an effective way to education and research, for in spite of exceptions the general pattern of our state universities is excellent. Nevertheless, federal and state support are markedly different matters, and alert understanding will be needed as the system of federal support develops.

A comparable danger—that there will be over-emphasis on the military aspects of science—is to the fore at the present time since federal funds are flowing from the military services into basic research. This system, however, is undoubtedly a temporary one, and the support is being rendered in a highly intelligent manner without undue control by officers of the military services and with strong emphasis on fundamental research. After the experience of the past six years, a tendency in this country toward over-emphasis on applications to military matters would perhaps be natural, but this has been to the present avoided so far as fundamental research is concerned. There is the converse danger, of course, that in a period of reaction such as we experienced between the two world wars, we might neglect the national security to our peril. The fraction of the country's effort in applied research and development which need be devoted directly to national security still remains to be determined; the proportion naturally depends upon the progress of world affairs. The maintaining of a just balance will not be easy.

Finally, there is a danger that the fierce present light shown by extraordinary application may blind men to the lasting cultural and philosophical significance of science and that therefore the moral and intellectual endorsement of this phase of the search for truth may diminish. Since indeed fundamental research will always be assured of financial support simply because it is the source of knowledge that can be applied indifferent ways, this hazard is not one of material impoverishment but rather one of intellectual if not spiritual disparagement. In this country in particular, the popularization of science has fed us with the spectacular until the public must be near satiation, and has already to a degree served to obscure from the less thoughtful the deeper significance of the exploration of the unknown—its relevance in all of man's most earnest ponderings upon himself, his fellows, his environment. Without doubt this serious interest in science—because of the basis for broad philosophical reasoning furnished by its advances—persists. It persists also because of mankind's innate curiosity, and a regard for the new and unique in thought. . . .

27

SHOULD SCIENTISTS RESIST MILITARY INTRUSION? (1947)

The military presented a unique challenge to the freedom of scientists. Committed to publishing their findings and openly sharing their work with peers around the world, scientists were uncomfortable with the military's preference that some research vital to new weapons and national security be performed in secret. In this comment to the American Scholar, *published by the Phi Beta Kappa Society, Bush elaborates on perspectives presented in the same issue by Dr. Louis Ridenour, a nuclear physicist who gained prominence for his role in developing combat applications for radar during World War II. After the war, Ridenour advised the new U.S. Air Force on research. In his comment, Bush insists that the tensions between the desire for open science and the imperatives of security in a hostile world can be balanced, "if we fully recognize" what's at stake in specific situations. Bush sought to reassure the public, as well as scientists and engineers working on classified military projects, that managing this tension is possible. But his optimism wasn't shared by all scientists.*

SHOULD SCIENTISTS RESIST MILITARY INTRUSION?

The long tradition of freedom of knowledge is characteristic of science to which Dr. [Louis] Ridenour properly gives much emphasis in his analysis of the attitudes which he holds scientists may take toward government support of research. Inevitably such emphasis tends to bring into disproportionate prominence the fact that some programs connected with the national defense must for the common good be conducted in secrecy, and to throw into correspondingly disproportionate obscurity the fact that—even under the temporary system whereby Federal money is channeled through the Armed Services—a very substantial part of the basic research thus financed is as free of restrictions as ever research has been. I quite agree with the general treatment of this important matter which Dr. Ridenour ably presents. The great point is not that there are dangers that research in this country may be placed on a false or wise basis (there have always been such dangers); the great point is that if we fully recognize them we can surely avoid them.

As for the possibility that continuing use of public funds to support research might weight the balance too heavily on the applied research side, I believe that availability of federal funds does not raise any new problem, although it certainly brings into relief an old one. If our scientific undertaking as a whole is to be of the maximum social value, effort must be properly distributed in the basic research which discovers new knowledge, and in the applied research which makes knowledge effective in controlling the environment and bettering the conditions of life. Scanting of either form through overemphasis on the other is obviously shortsighted. The interest of the scientist and the interest of the public are parallel here. During the wartime emergency, great emphasis on applied research was natural and proper. It should be noted, however, that that emphasis was a result of the war needs, not of the federal financing; it does not at all follow that federal financing in times of peace means a continuation of extra demands for applied research.

The greater significance of both the present interim function which the Services are ably performing, and the hard thinking which is being done on the formulation of legislation for a National Research Foundation, lies in two facts that are too often overlooked. These facts are that we are now in the process of recognizing, defining and establishing a clear relationship between government and science as one of the major

American Scholar Forum on question of "Should scientists resist military intrusion?," *American Scholar*, Spring 1947.

components of our civilization; and that this time we are carrying that process through with more consciousness of what we are about than we had, say, in developing a similar relationship between government and such other components of society as agriculture or the transportation industry. The phenomenon should be of profound interest to social scientists, for it seems to indicate not only that research—which we may here consider to be the deliberate application of intelligence to the increase of knowledge and to the control of the environment—is now recognized as a major concern of the people as a whole, but also that we as a nation have attained maturity in this regard.

28

SCIENCE, DEMOCRACY, AND WAR (1949)

As competition with the Soviet Union intensified, Bush addressed the fear of many Americans that dictatorships might produce faster techno-scientific innovations than democracies. Bush's book, Modern Arms and Free Men, *from which this essay is excerpted, was published to acclaim partly because of timing. As the book went to press in late September 1949, Americans received the shocking confirmation that the Soviet Union had exploded an atomic bomb on August 29, 1949. Bush helped to verify, at the request of President Truman, that the Soviets had indeed done so, and barely more than four years after the United States. The Soviet achievement, which Bush had predicted in a memo to Roosevelt in 1944 (see chapter 18), heightened worries among influential Americans that freedom and democracy might not be essential to producing distinctive feats of innovation. The Soviets were, despite the equation of scientific achievement with political freedom, capable of matching the Americans. In* Modern Arms and Free Men, *Bush insists that democratic approaches to supporting and directing scientific and technological research and development will ultimately triumph over totalitarian states such as the Soviet Union. As Bush wrote in a hastily written foreword to* Modern Arms, *dated September 26, 1949, the shock of the Soviet achievement underscored the importance of viewing "the principles that underlie our democratic system" as durable; these principles, he reminded readers, "are not dependent upon the quickly changing events of day to day." In short, Bush wants his audience to ask tough questions about the risks of a growing rivalry between the United States and the Soviet Union and to think hard about "how to avoid another war."*

SCIENCE, DEMOCRACY, AND WAR

One question is in the mind of every American as he faces the blurred future: Will the coming generation of our youth have to fight in another desperate war?

All our opinions, all our acts, are conditioned by this question. We tax ourselves heavily, and more or less cheerfully, in the belief that generous resources will help. We extend aid to Europe, at heavy cost in materials and labor, even at peril to our economy, in order that this aid may help rebuild a bulwark against war. We instituted a peacetime draft, and we strive to unify our military organization, in the hope that if we are fully ready we will not be forced into war.

The central question breaks down into many. Is it true that a new all-out war, with atom bombs and biological warfare, would destroy civilization and drive us back to the dark ages? Is the case so desperate that a prophylactic war is justified in order that we might at least meet the inevitable at our own time and on our own terms? Can a democratic regime develop great military strength without distorting its true nature? Has the time not come when the peoples of the world, in terror before the threat of war, will build one world under law? Or is peace so sweet to those who live at the edge of the abyss as to be bought at the price of chains? What can democracy offer to those in distress toward building a world in which free men will live in concord, peace and understanding?

There are no precise answers to these questions, just as there is no complete answer to the bigger problem of how to avoid another war. There are powerful factors present: science and democracy. Modern science has

---

From chapter 1, "Science, Democracy and War," excerpted from Vannevar Bush, *Modern Arms and Free Men: A Discussion of the Role of Science in Preserving Democracy* (New York: Simon & Schuster, 1949). The book, chosen as a featured selection by the influential Book of the Month Club and a neglected classic in the field of war and technology, has been out of print for more than fifty years.

utterly changed the nature of war and is still changing it. And the democratic process has given us new controls over our destinies that are subtle, only partly understood and also changing. This book [*Modern Arms and Free Men*] is an examination of the vast process of change in which science and democracy are both affecting the nature of war.

I know no scientific formula with which to explain this great and intricate interaction of old and new forces. Its order of complexity is too high and it has too many variables for either a human or an electronic brain. But it is a process that we can begin to understand, better at least than any generation before ours, and with understanding comes control. For ten years, thanks to the accidents that direct men's lives in a democracy, I was in a position to see as much as any single man could see what science has done and can still do to the art of warfare. It is part of the obligation of any citizen who has been given such responsibility and opportunity as I have, no matter by what accident, to set down the record for what he has learned, and to share with others any light it may throw on the great question of war or peace that haunts us all.

I am not much of a prophet; there is a great deal of guesswork inevitably involved when we attempt to predict just what applied science may still do to our lives. Yet I have specialized in the development of new weapons and wrestled with the scientific problems of total war. I have also watched with fascination the ponderous turning of the wheels of government and the weaving of men's relationships with each other into patterns of incredible complexity. I have evidence that supports the two chief conclusions of this book. I believe, first, that the technological future is far less dreadful and frightening than many of us have been led to believe, and that the hopeful aspects of modern applied science outweigh by a heavy margin its threat to our civilization. I believe, second, that the democratic process is itself an asset with which, if we can find the enthusiasm and the skill to use it and the faith to make it strong, we can build a world in which all men can live in prosperity and peace.

This old challenge has become a new one as a result of the application of science to war in a degree that has completely altered warfare. The combination of science, engineering, industry and organization during the last decade created a new framework that rendered conventional military practice obsolete. Radar, jet aircraft, guided missiles, atomic bombs and proximity fuses appeared while we were fighting. They determined the outcome of battles and campaigns, even though their determining nature was not fully exploited in that contest. Over the horizon now loom radiological and biological warfare, new kinds of ships and planes, an utterly new concept of what might be the result if great nations again flew at each other's throats. It is this which makes the thinking hard.

We cannot take refuge in the assertion that these are matters for specialists: for a State Department to carry on a new form of international discussion, for a Defense Department to prepare us for a new kind of war. We are not a dictatorship, where a single distorted mind names the tune and all the lackeys dance to it. We are a free people, and as we think, so will our public servants act.

Must every citizen, then, grasp the full nature of atomic energy, evaluate the modern submarine, predict the consequences of supersonic flight or grasp the mentality of those who rave at us in councils? It is absurd, of course. But every citizen, in a strange subtle way, visualizes where we are and where he feels we are going, and from this is distilled, in a way we hardly understand, our national policy in every regard. It is not just through votes, or editorials, or commentators that this action occurs, but more indirectly and through a thousand channels.

Since the [second world] war ended [in 1945], this elusive and powerful force, this mass concept, this public opinion, has ruled that we should enter wholeheartedly, in spite of irritations and annoyances, into the attempt to build some sort of United Nations. It has ruled that we should build a strong military machine and have it ready. It has stiffened our backs and frowned on exposure of weakness or over-readiness to compromise. It has most certainly rejected any idea that we should become a conquering nation, or strike early ourselves in the attempt to avert a later and a more desperate war. It insists that we control the traitors in our midst and somewhat bewilderingly that we do not sacrifice our essential freedoms in the process. It has even begun to insist that special and selfish interests be regulated with a compass that will not wreck the national strength we need. It has placed us on the path we now pursue, yet does not know exactly where it leads.

How has public opinion done all this? In discussion and criticism our people somehow sense out the big issues, of course. But public opinion does this principally because men understand men, and, through some process that is still mysterious, select those to be trusted, which is the essence of the democratic process. From the whole seething fracas emerge a national attitude and policy which become the guide for all who manage affairs in the public interest. There is a chain of trust. We do not elect a President because we think he understands atomic energy in all its ramifications; we know and he knows he does not. In November 1948, we elected a President in a close contest where two opponents [Harry Truman and Thomas Dewey] were selected by a system that, illogical as it is, nevertheless produced contestants who truly appealed to us, and we elected one because the majority preferred him. The issues were stated and argued, but did not really determine the outcome. This was determined by the millions and their ballots on the basis of where they preferred to place their destiny. The uppermost thought was whether he who was chosen [Truman] would lead and judge better, as compared to his defeated opponent, in the interest of the common citizen, marshalling about him those who are specialists, in science or war making, or diplomacy, or politics, marshalling them with a stout heart and common-sense judgement— whether he too could truly judge whom he could wisely trust.

It is highly important that the general outlook of the people be sound as we face the future. If we had been in abject terror, facing a new inevitable war that would destroy our cities, our farms and our way of life, we would have followed some Pied Piper in the last election who would have led us into the sea. This we emphatically did not do. In spite of alarms, in spite of the prophets of doom, we face the future with resolution. If as a people we had felt all-powerful, that we could speak and the world would tremble, that we had a mission to rule the unenlightened, that we were a super race, we would have followed a demagogue. There was not even a single demagogue of the sort in sight on the national horizon. The steadiness of purpose of the American people is our hope and refuge.

This national attitude is extracted from a maze of conflicting argument centering about two focuses: the nature of the democratic process and what the applications of science hold in store for us all. The two are intimately intertwined, for science does not operate in a vacuum, but is conditioned by the political system that controls its operations and applications. The discussions on the air or about the stove at the corner store revolve about these two central subjects. They are not always recognized as being present, for the talk may be on the next crop or the ambitions of the local sheriff, but they are in the background. For they determine our destiny, and we well know it. If the democratic process will work to foster an effective government, if science will cure our ills and not merely provide the means by which an aggressor can suddenly destroy us, we have a rosy outlook and can quarrel about minor things without fear. If we are headed over a cliff, with our means of progress out of control; if our form of government is transitory and bound to transform itself into selfish rule by some dominant group; if the application of science has finally doomed us all to die in a holocaust, there is little use in arguing about the drought or the next [labor] strike. The two central matters are interconnected. What science produces, in the way of applications within its own changing limitations, depends upon what is desired by authority, by those who rule or represent a people. Pure science may go its own way, if it is allowed to do so, exploring the unknown with no thought other than to expand the boundaries of fundamental knowledge. But applied science, the intricate process by which new knowledge becomes utilized by the forces of engineering and industry, pursues the path pointed out to it by authority. In a free country, in a democracy, this is the path that public opinion wishes to have pursued, whether it lead to new cures for man's ills, or new sources of a raised standard of living, or new ways of waging war. In a dictatorship the path is the one that is dictated, whether the dictator be an individual or a self-perpetuating group.

When for the first time in history the decision was taken to recognize scientists as more than mere consultants to fighting men, I was living in Washington, and President Roosevelt called on me to head the job. Abraham Lincoln had set up the National Academy of Sciences during the Civil War, and Woodrow Wilson had authorized the National Research Council during the first world war. Both did good work, and their histories are illustrious, but neither was given large funds or authority. In the National Defense Research Committee and the Office of Scientific Research and Development, in the second world war, scientists became full and responsible partners for the first time in the conduct of war.

We had, during the war, approximately thirty thousand men engaged in the innumerable teams of scientists and engineers who were working on new weapons and new medicine. We gathered the best team of hard-working and devoted men ever brought together, in my opinion, for such a task. We spent half a billion dollars. Congress gave us appropriations in lump sums and trusted us to decide on what projects to spend the money. In uniform but without insignia, some of our men were on the battlefields and in the planes and ships in every theater of the war.

At the same time, I watched a great democracy bend itself around this new development and give it life and meaning. In this I was a rank amateur, knowing little about the details of the democratic process but believing in it. I saw it work. We contested with generals and admirals, but the new weapons were produced and used, and we wound up friends. We argued among ourselves, but of roughly thirty-five men in the senior positions in the scientific war effort only one man was absent when the war ended, and this was because of illness. We did a job that required fantastic secrecy, and yet we won and held the confidence and support of the military, the Congress and the American people. We were a varied group with all sorts of backgrounds and prejudices, and yet we developed a team technique for pooling knowledge that worked.

This is not a history of what science did in the war; that has already been written. It is an attempt to explore its meaning in the relations between man and man, as individuals and in the organizations they create. Since the beginning of organizations there have been two controlling motivations that have held them together. One is fear, utilized in the elaboration of systems of discipline and taboos. The other is the confidence of one man in another, confidence in his integrity, confidence that he is governed by a moral code transcending expediency. Most governing organizations have involved a mixture of these motivations; they always will as long as the nature of man remains unaltered, but one may be controlling and the other subsidiary, incidental or extraneous. There has been a general feeling that the second is the higher motivation, but that it is inherently weaker in dealing with the harsh and complex conditions of existence.

The subject is of extreme seriousness to us today. The world is split into two parts that confront each other across a gulf. On one side is a rigid totalitarian regime, ruled through fear by a tight oligarchy, which sees only two possibilities: it will conquer the world or succumb in a final struggle. On the other side is a diverse group of nations, with democracy as the central theme, which aspires to a world of peace under law and would bring it about by advancing collaboration and mutual confidence, respect for the given word, integrity, as higher and essential bases of action if the world is to be more than a mere police state. Neither is absolute in its nature. Within the totalitarian regime there is still an aspiration for freedom; there is, moreover, in the great mass of those rigidly controlled from above an idealism, a neighborly helpfulness, a grasp of something higher than selfish ambition, which still persists in spite of regimentation, propaganda, and the evils of the secret police. Within the democracies there is still plenty of chicanery, a negation of principles in the treatment of minorities, abuse of the necessary police power. Yet the issue is still clear. On the one hand is an absolute state, holding its people in subjection and molding them for conquest by force or trickery. On the other, there is hope for better things.

Through the pattern of modern thinking runs a doubt, a question as to whether a system based on the dignity of man, built on good will, can be sufficiently strong to prevail. The thesis of this book is that such a system is far stronger, in dealing with the intricate maze of affairs that the applications of science have so greatly elaborated, than any dictatorship. The democratic system, in which the state is truly responsive to the will of the people, in which freedom and individuality are preserved, will prevail, in the long run, for it is not only the best system, the most worthy of allegiance that the mind of man has built; it is the strongest system in a harsh contest.

The striking success of the application of science in controlling nature for our purposes has not only modified the conditions of the contest, producing radio for propaganda in a cold war, and atom bombs to consider in estimating the consequences of allowing it to become hot; it has also modified the underlying philosophy with which men approach the problems of their organization and government and every other aspect of their existence. Totalitarianism and tyranny, and the struggle of men for freedom, existed long before science became widely applied. But the success of science has given concrete form to the clash of philosophies that now divides the world. On the one hand, fear is seized upon as the only dependable motivation, and moral codes are discarded for a blatant expediency, because of a crass materialism that has become embedded in a totalitarian regime, and this regime enthrones science as its model. On the other hand, there is faith, even though at times it may not extend beyond faith in the dignity of man.

The philosophy that men live by determines the form in which their governments will be molded. Upon the form of their government depends their progress in utilizing the applications of science to raise their standards of living and in building their strength for possible war. Upon this form depends the effectiveness with which they can provide for security against the ravages of nature and of man. Upon the form depends also their progress in securing justice and maintaining opportunity. Upon it depends the outcome as to whether life is worth living.

Through it all runs the thread of the impact of science in altering the world and the relations of men herein. It will not be enough to trace the current and future development of weapons or even the ways in which science may further alter our material affairs. We need to delve deeper, and we shall.

29

HOW SCIENCE WORKS, OR DOESN'T, UNDER TOTALITARIANISM (1949)

In this second selection from Modern Arms and Free Men, *Bush inquires more deeply into how science works, and doesn't work, under totalitarian regimes, against the backdrop of a deepening Cold War between the U.S. and the Soviet Union. He argues (hopefully and honestly, if not always convincingly) that the core traits of "the Communist state"—rigidity, intolerance of dissent and resistance to open flow of ideas and facts—"are fatal to true progress in fundamental science." The relevance today for Bush's argument can be found in debates over whether the People's Republic of China can sustain technological creativity and scientific advance in a dictatorship that permits economic freedom but not political freedom.*

HOW SCIENCE WORKS, OR DOESN'T, UNDER TOTALITARIANISM

We cannot here [decide] whether the [Communist] system contains the seeds of its own disruption, whether its satellites can acquire the strength to defy it, whether it will evolve into a new pattern of despotism or whether, indeed, as we hope, it may ultimately come to rely more on persuasion, less on force and so become what we consider a more democratic system. [I'm] concerned with the influence of science upon war and the interrelations of science and government where the potential pressure of war is a determining factor in our future. We can examine briefly the way in which science will operate within such a dictatorship and there is no surer index of the weakness of the Communist state than the one afforded by that examination.

The weakness of the Communist state resides in its rigidity, in the fact that it cannot tolerate heresy, and in the fact that it cannot allow its iron curtain to be fully penetrated. All these things, vital to totalitarianism whether right or left, are fatal to true progress in fundamental science. They are not nearly so fatal to the application of science, but they are a severe deterrent to even the healthy growth of this along novel lines.

Dictatorship can tolerate no real independence of thought and expression. Its control depends entirely upon expressed adherence by all to a rigid formula, the party line. Its secret police must be ever alert to purge those who would depart from discipline and think their own thoughts, for departure would soon lead to a vast congeries of independent groups defying central authority, and the system would break.

No true art, no true fundamental science, can flourish long under such a system, no matter what the individual genius may be. Musicians, some of the finest and most creative in the world, are disciplined because a commissar does not like their music, and they bow to the inevitable, apologize and admit their errors and promise to conform. A great scientist is torn from his post and sent into cold exile because he dares assert that there is validity in the modern theories of genetics, contrary to the state teaching that environment is all-controlling, and he is replaced by a charlatan who will see to it that the state theory is taught to young scientific disciples and that all the research is based on a blatant fallacy and an unsupported hypothesis.

Under such a system art will eventually become merely a dull adjunct to monotonous propaganda. Science will eventually become a collection of superstitions and folklore. Men of genius will languish and succumb to

Excerpted from chapter 14, "Totalitarianism and Dictatorship," from Vannevar Bush, *Modern Arms and Free Men: A Discussion of the Role of Science in Preserving Democracy* (New York: Simon & Schuster, 1949) (1949).

discouragement. It will take time. The spirit of a great people that has produced sparkling figures in the past in music, in mathematics, in art and science, will not be broken in an instant or a single generation. The spirit of genius is strong, and it has always risen through obstacles to make its presence known. But in the long run, that very science, on the distorted interpretation of which is built the philosophy of the group that rules in Russia, will become a sham and a mockery.

The situation is not nearly so clear when we come to the application of science, to that interplay between scientists and engineers, industry and government, by which the fruits of new advances in fundamental science are made available for practical use in manufacture or agriculture or war. To throw light on that matter we shall digress to chart illustrations from history in the field of atomic energy and in applied science.

In the years since the war, the honest and determined attempt of this country to place the atomic bomb and all its works under genuine international control has failed. Not quite all of the world was ready for it. It was a sound move; it will stand in history as the wisest step ever advanced by a great nation in its full strength and in the flush of victory. It was well thought out, to avoid the pitfall of false reliance on ineffective controls, and it expressed clear grasp of the practical relationship of international-control activities in a world still steeped in intense nationalism. It remains in abeyance and awaits a better day.

[International control of atomic weapons] failed because an essential element of any system of control that is not to be a delusion and a snare is an international system of inspection that works. The rulers of Russia could not accept such a system for it meant the penetration of the iron curtain in no uncertain manner. Once that curtain was pierced, so that international committees could move freely about within the country, consulting with those who operated factories and laboratories, examining into the flow of raw materials and finished fission products and checking against their diversion, uncontrolled by police as long as they adhered to their proper affairs, the door would soon open wide. The population would learn the true state of affairs in the world. Disaffected citizens would find means to cross borders, their families with them, and escape the clutches of the secret police. Industrial units would not be completely subservient to central control. Controversy, departure from the party line, would spread, rigid discipline would be lost.

Those who would rule Russia would not take the risk—even though it was probably desirable from the Russian standpoint, even in a strict military sense, that the atomic bomb be somehow removed from the armament of nations. There may have been a variety of reasons why the rulers of the Soviet Union would not take the risk. One of the reasons may well have been that the rulers of Russia did not want to follow the suggestion that they personally commit suicide for one does not lose a high post there and retire—he loses the post and dies. The chain of events set in motion by true international inspection would probably, sooner or later, have challenged the absolute power of the central dominating group. So the atomic-energy proposal, and all proposals for controlled armament failed, and we became committed to an armament race in spite of ourselves.

But—and here is the weakness of rigidity—the iron curtain operates in both ways, to keep people in and to keep ideas out, and among information and opinion it excludes there will be much that is needed for great progress, not only in fundamental science but also in its applications. Scientific publications of all sorts, pure and applied, will cross the border, but this is not enough. Unless they are accompanied by free analysis and discussion that cuts across international borders they will lose much of their value. As the Soviet Union places complete bars upon effective international scientific exchange, and to the extent that it regiments scientific discussion within its own borders, it places upon itself a substantial handicap in any technical race, a greater handicap than is readily appreciated by us who live in freedom. . . .

Yet even more than unrestrained criticism is needed for healthy advance in fields that involve many skills and techniques. Interchange and collaboration on a give-and-take basis are also needed. The larger and more complex the effort, the more this becomes essential to progress. We return to the history of atomic energy for an illustration.

The atomic-energy program during the war was to the nth degree the sort of collaborative program that was impossible at that speed outside democracy. It was not merely a matter of new physics and its incidental application—very far from it. True, some of the finest theory and experiment in the physics of the atom was involved, calling for ingenuity and resourcefulness, mathematics of a higher order, and judgement such as can be exercised only by men who are utter masters of their craft, and all this was performed magnificently by the physicists of this country, England and Canada, in close interchange and at unprecedented speed as the program of application proceeded. But then came the heaviest part of the job. It involved new, dangerous and complex chemistry, the most refined sort of chemical engineering [and an] industrial organization that tied together effectively the performance of ten thousand firms which supplied parts, built new and unheard-of devices, constructed and operated enormous plants where the whole affair functioned as an interlocked unit. It involved joint action of diverse groups, theorists, engineers, instrumentalists [and] designers, in the production of fissionable materials and in the construction of the bomb itself. It involved management that reached a new order of functioning to bring all these elements together in an intense race against time, where nerves were bound to be frayed and patience short. It involved collaboration between military and civilian organizations, with their widely different approaches to organizational rules and systems. It required integration of new elements into the strange structure of government and competition with other programs of highest priority in the maelstrom of war.

The keynote of all this effort was that it was on an essentially democratic basis, in spite of the necessary and at times absurd restrictions of secrecy and the formality that tends to freeze any military or for that matter governmental operation of great magnitude. If certain physicists thought the organization was functioning badly in certain respects, they could walk in on the civilian who headed that aspect of the effort and tell him so in no uncertain terms. They not only could, they most certainly did; and the point is there was no rancor and old friendships were not destroyed in the process. If civilians and military disagreed, as they often did, there were tables about which they could gather and argue it out. Punches did not need to be pulled, and no one kept glancing over his shoulder. If there were international misunderstandings between allies, and there were, they could be frankly discussed—sometimes with more heat than light—but always with a prevailing atmosphere of genuine desire to arrive at the conclusion that made sense and that best got on with the war. If a young scientist had an idea he did not have to pass it through a dozen formal echelons and wait a year; he could talk it over with his fellows and with superiors of accepted eminence in his own field and be sure it would be weighed with unbiased judgement by men of competence. The system worked and it produced results.

Now the Nazis, the totalitarians of the right, were also trying to make a bomb, and they failed miserably. They had the same opportunity that we did; the starting gun for the race went off with the experimental confirmation of the phenomenon of fission in 1939, and this was known all over the world. At the end of the war it was found that they had made little progress. They had not accomplished 5 percent of the task that was successfully brought to a conclusion in this country, with the collaboration of England and Canada. That they were far behind in the race was not known until the Alsos mission [an intelligence-gathering operation by the Allies in Europe to learn firsthand how close Germany was to developing an atomic weapon] revealed the true state of affairs after the fall of Stuttgart [in 1945]. Until then we felt they were close competitors and even that they might be six months ahead of us, which would have been disastrous. We maintained essential secrecy well in this country, and gave the Nazis credit for being equally adroit so that their possible progress was always a burr under our saddle until thorough intelligence work revealed the truth, fortunately in time, so that the European campaign could be carried through to its finish in the light of accurate knowledge of an important consideration.

Why were they so far behind? Bombing and the destruction of needed industrial facilities account for some of the lag. Limited availability of critical materials accounts for some. But the real reason is that they were regimented in a totalitarian regime. There was nothing much wrong with their physicists; they still had some able men in this field in spite of their insane rape of their own universities. They were not as able as they thought they were, or as they probably still think, for their particular variety of conceit is incurable. But they were able enough to have made far greater progress than they did. Their industry certainly demonstrated that it could produce under great stress such complicated achievements of science and engineering as the jet plane. Their Fuhrer and their military were certainly keen for new weapons, especially a terror weapon with which to smite England. Yet they hardly got off the starting line on the atom bomb.

A perusal of the account of German war organization shows the reason. That organization was an abortion and a caricature. Parallel agencies were given overlapping power, stole one another's materials and men and jockeyed for position by all the arts of palace intrigue. Nincompoops with chests full of medals, adept at those arts, presided over organizations concerning whose affairs they were morons. Communications between scientists and the military were highly formal, at arm's length, at the highest echelons only, and scientists were banned from all real military knowledge and participation. Undoubtedly the young physicist who penetrated to the august presence of Herr Doktor Geheimrat Professor said *ja* emphatically and bowed himself out, if he did not actually suck air through his teeth. The whole affair was shot through with suspicion, intrigue, arbitrary power [and] formalism as will be all systems that depend for their form and functioning upon the nod of a dictator. [The Germans] did not get to first base in the attempt to make an atom bomb.

This all became clear when research teams were sent into Germany after the war to study such problems as the German handling of strategic materials, manpower, propaganda, military organization and co-ordination. All of these teams were amazed. The Germans had made astonishing mistakes in every sphere of action, which continued throughout the war without correction. They made thousands of little mistakes as well as the big ones.

It had been popular with us for many years to complain about the wastefulness of the democratic system of government. We had come to expect 50 cents worth of results for every dollar paid in taxes. Benevolent dictatorship was thought to be efficient by comparison, and it was held that a dictator could get results much more generally and cheaply. What we found at the end of the war exploded this myth for all time. Every team we sent returned convinced that the democratic system is clearly more efficient, dollar for dollar and hour by hour, than any totalitarian system. The criticism applies to the German system before the Nazis took over for it was autocratic long before that with only a brief interregnum of attempted democracy, which failed because the people were not genuinely seeking freedom. The Nazis merely developed and tightened the system they found in existence.

The teams also found the key to the difference, the key to the effectiveness of the democratic system. It lies in the fact that in a democracy criticism flows both ways, up as well as down, and we shall have more to consider on this point later. Here it may be merely noted that this factor, plus democracy's ability to call to the aid of government at need the views and judgements of experts in any field who operate, when thus called, with complete frankness, far more than offsets the apparent looseness of democracy and renders the whole structure live and virile, changing and adaptable, rather than frozen into a pattern where any absurdity or any incompetent can persist if politically entrenched.

The type of pyramidal totalitarian regime that the Communists have centered in Moscow is an exceedingly powerful agency for cold war. It is capable of holding great masses of people in subjection, indoctrinating them in its tenets and marshalling them against the free world. It can force its people to enormous sacrifice and thus build great quantities of materials of war. It can educate large numbers of men and women in science and engineering, construct far-flung institutes, mechanize agriculture and ultimately create mass production of the manifold things it needs. But it is not adapted for effective performance in pioneering fields, either in basic science or in involved and novel applications. It has many of the faults of the German dictatorship, magnified to the nth degree. Hence it is likely to produce great mistakes and great abortions.

This does not mean that we can flatly disregard the Communist state and cease our advances in the techniques of war. The Communists can copy and improve and a whole mass of scarcely developed techniques remains from the last war as material for this process. It does mean that we must continue to break new ground, and that we can do so with our heads high, for we have a system essentially adapted for the purpose, if we do not distort it or sacrifice it to false gods of fancied efficiency.

The Russian people are rugged, long-suffering, tough. They are likeable. They have among them men of imagination capable of giving much to the world. They are capable of going far even under dictatorship. Should they escape from it by some miracle, should they become in some manner a part of the free world, they could be a great people, with an important share in some ultimate community of nations. In the meantime, as we contest doggedly with the system that controls them, and hold its power in mingled fear and contempt, we should not mistake the super-imposed hierarchy for the people themselves. The sound way in which to carry on the struggle is to preserve without question a system that is better, better even for making war, and far better for advancing science and gathering its fruits, and we should let the world, and the people of Russia when we can, know about it in its genuine characteristics and form.

As we do so, as we conduct the disagreeable contest of the cold war at enormous disadvantage, we may be reassured. The system with which we contend cannot possibly advance science with full effectiveness; it cannot even apply science to war in the forms it will take in the future, without mistakes and waste and delay. Moreover it cannot possibly alter its pattern and become fully effective without at the same time becoming free, and if it becomes free the contest is ended.

30

THE ESSENCE OF SECURITY (1949)

In the early years of the Cold War, Bush often expressed views on national security, a growing preoccupation of politicians, soldiers, and civic leaders. He often served as the public face of a powerful faction of "establishment" scientists who sought to guide, or manage, the marriage between science and government, soldiers and technologists. In his role as a science "statesmen," Bush frequently reminded younger scientists and engineers of the importance of contributing to national security by applying their knowledge and skills to emerging international threats. "Now I am no pessimist. I believe in the democratic system, and I believe in the sound common sense of the American people," he insists in this address to students and faculty of his beloved MIT. While identifying the limits of atomic weapons as well as their fearsome strengths, Bush highlights the ideals of freedom and obligation which he believes will sustain a civic-minded techno-scientific community in the United States. Speaking directly to a new crop of talented students, Bush declares, "you men of influence in the coming generation will determine where we come out," not only in competition with the Soviet Union but as leaders of a free society and freer world.

THE ESSENCE OF SECURITY

The fundamental question of our times [is] the strength of the free demo-cratic system as compared with the strength of the totalitarian system, and the ability of each system to grow in strength in the years to come. Until such time as reason and good will prevail, and a rational answer to the prob-lem of live-and-let-live in a complex world is established, this comparative estimate will stand as the most imponderable in the world's vocabulary. By and large, all questions of policy—national or international, economic or political, immediate or long-range—take on importance and urgency to the degree—and only to the degree—that they bear on this basic issue.

Russia is a closely controlled dictatorship, a police state, with full ulti-mate management of the details of the life of every citizen. [The Russian government] can hold, and has held, the standard of living down to a small fraction of ours, denying its people the simple comforts of existence, in order to focus effort on guns and atoms. In the long run a totalitarian state cannot compete with a free people in the advancement of science, for dictation and dogma are contrary to the free spirit of inquiry, which is the heart's blood of scientific advance. But, in the short run, it can produce what it wishes to produce, and ignore the sufferings of its people, up to a limit (and that limit is high). It can produce an atom bomb, and has done so. In time, how much time is arguable and important, [Russia] can pro-duce a stock of atomic bombs.

The atomic bomb does not stand alone. It is not an absolute weapon. It is part of a vast and intricate armament, and much of the nature of that armament was spread out for all the world to see in World War II, and is known to many technicians in Russia and elsewhere. Russia can build fleets of bombers, jet aircraft, radar networks as well as guided missiles.

But can [Russia] build all these at once in ample quantities, in reli-able form, operated by well-trained men of initiative and resourcefulness? The answer to this question depends upon what one believes the Russian economic system can stand without collapse. But, whatever the estimate, there is no doubt that Russia can build, and is building, a formidable mil-itary machine. Whether it does so from genuine fear of attack, or in a

---

The text is from Bush's address to MIT's student body convocation, December 5, 1949. Reprinted in the January 1950 issue of MIT's *Technology Review* magazine.

dream of conquest, is beside the question; we face the fact that, barring an improbable early internal collapse, such armament will appear in the hands of an absolute closely knit central governing group of men who distrust us and would destroy us if opportunity offered.

We can meet that threat if we are strong. We can in fact meet it without war for those in the Kremlin recognize strength if they recognize nothing else. The fact that we can meet it was recently proved when this country, bringing to bear on grim business the resourcefulness and initiative nurtured in freedom, armed its allies and joined with them to strike down the Nazi might.

But we cannot meet [the Russian challenge] if we turn this country into a wishy-washy imitation of totalitarianism, where every man's hand is out for pablum and virile creativeness has given place to the patronizing favor of swollen bureaucracy. Dictatorships can compete with dictatorships, and free virile democracy can outpace any such in the long pull. But a people bent on a soft security, surrendering their birthright of individual self-reliance for favors, voting themselves into Eden from a supposedly inexhaustible public purse, supporting everyone by soaking a fast-disappearing rich, scrambling for subsidy, learning the arts of political logrolling and forgetting the rugged virtues of the pioneer, will not measure up to the competition with a tough dictatorship. If we go all the way down the path to dependence and render ourselves a people fawning for handouts on an intriguing bureaucracy, Russia can cease its building of war machines. It will conquer the world without them.

Now I am no pessimist. I believe in the democratic system, and I believe in the sound common sense of the American people. Moreover I believe thoroughly that we have the wit to recognize a dangerous trend, reverse it before it's too late, and laugh at sirens with crack-brained economic theories who would guide us down an easy path over a precipice. I believe also that the past generation, with a rise in the power of the common voter, and an increase also in his perception and grasp of public affairs, has brought with it highly salutary progress toward protection of the small man against the hazards of nature or of his grasping neighbor. We live in a better world because of awakened public consciousness of its power and possibilities. We have done much for the underprivileged and more for the laborer at the bench, and it is well. But we can outpace ourselves, attempt too much and wreck the industry on which all material progress depends. Still I believe the American people are too tough-minded to pursue a will-o'-the-wisp over a cliff. I know they can add, and I do not believe they are fooled by stories of magic wands or of inexhaustible treasures. If I am wrong, we are in for disaster.

But my main point today is to ask what you men propose to do about it. Some of you will differ with my point of view completely, but no matter. I ask what you propose to do about your own theory of the dilemma and its solution. Unless you take the naive point of view that the building of a powerful military machine overseas is no concern of ours (and that delightful fiction seems to have disappeared even from the ranks of the isolationists) there is something to be done. Unless you take the equally naive point of view that this is an affair for our own military men and no concern of the private citizen, you have a problem squarely on your doorstep. Moreover it is peculiarly yours—for the strength of this country means more than arms and services, it involves the entire strength and prosperity of the country, which is a matter directly in your area. The question is whether you will be content to build in a technical sense, and let someone else worry about the larger problems.

Some of you, unhappily, will spend your time and effort exclusively in passing technical courses, and thoroughly shun those dealing with man's relation to man. You will not pay the slightest attention to the political maelstrom about you, and look down on those who take an interest in politics or who have the hardihood actually to practice it. You will regard such things as history, economics, mass psychology [and] foreign relations as soft generalizations not worthy of the steel of a man who can manipulate a Fourier integral. Or at the other extreme some will browse in generalities and amuse yourselves with vague non-rigorous speculations and know nothing of a Fourier integral whatever. You will read the newspapers and the current magazines—for relaxation. When you graduate with the M.I.T. accolade, you will seek out a post that pays a salary and affords the maximum of security, where promotion is sure and slow, and where no one ever got fired who didn't criticize the management unduly. I am not talking to the small fraction who will thus dodge the issue.

I am talking to those who will be men of influence, now and from here out, who have a feeling for the game and a will to live something besides the life of the oyster. I speak to those who are neither terrified by life nor lured into wishful thinking regarding a millennium. There are many such among you, as the history of this institution has proved, who would grapple with a tough world, and learn the subtle arts of doing so well. There are many among you who are men of influence today and who are preparing to be men of great influence tomorrow among your fellows, because of your grasp, your courage, your mental power, and your integrity.

There is an unfortunate tendency in this country to separate the great and small issues; to take the point of view that the large ones are determined in Congress and the Executive and that they are no direct concern of the individual citizen, pursuing his ordinary affairs within close horizons. Nothing could be further from the truth. The characteristics of the federal government—the dignity or lack of it with which we conduct our affairs, internally and in the complex relations between nations, the selflessness, or its opposite, of our governors throughout our far-flung government machine—are but the reflection of the character of the people. As the people think, as they demand effectiveness or tolerate demagoguery, so will our status as a nation evolve. The smallest act of a citizen, influencing his friends and neighbors toward sanity and far-reaching wisdom, multiplied a million times throughout the breadth of this land, can determine that the country as a whole will rise in its dignity and strength to live in prosperity and peace, and not succumb in a welter of petty selfishness and confusion. This is the essence of security.

You men, with the enormous privilege of higher training, in an institution of world renown—and thousands like you throughout this country—you men of influence in the coming generation—will determine where we come out. You will specify, you men who accept the challenge of life, whether we make of this great country of ours a country that grows in freedom and strength, and whether in that strength it will lead the world out of a morass. No one of you will determine it all or even a great part of it, for the world is large and complex and the efforts of no single man reach far unless he be singularly favored by chance and endowment. But each of you who wishes will shift the trends a bit, and is shifting them now; and together, you of youth who inherit the world, you will determine the outcome. I shall probably not be around to witness the full result, but I wish you luck.

31

THE ATOMIC BOMB AND THE DEFENSE OF
THE FREE WORLD (1951)

Bush was among the first Americans to think seriously about the military, political, and societal implications of nuclear weapons. By the early 1950s, he began to envision potential security benefits from a nuclear stalemate with Russia, notably the likely unwillingness of either side to engage in total war. He spoke with the confidence and authority arising from his central role in the American victory in World War II, the Manhattan Project, and as the first nuclear adviser to President Roosevelt and, later, President Truman. In the eyes of the public, Bush possessed a special knowledge of weapons of mass destruction and of the logic of Cold War with the Soviet Union. The text of this talk by Bush, delivered to a national radio audience of "fellow citizens" via the Mutual Broadcasting radio network, launched a series of weekly broadcasts aimed at educating Americans on the Soviet threat and the likelihood a "stalemate" achieved through nuclear deterrence; because, he argues, for Russia "to attack us would be suicidal." While president of the Carnegie Institution of Washington, Bush spoke to the nation as a representative, or voice, of the recently formed Committee on the Present Danger, a civic group of "wise men" (also known as the "Eastern Establishment"), who were trying to whip up support for the "containment" policies of President Truman. The appeal relied on fear of the Soviet threat and the presumption that the world was divided between free and unfree peoples and nations. The Committee, which Bush described as "non-partisan," was formed in December 1950 by Bush, James Conant, and Tracy Voorhees, a former under secretary of the army. Conant was the first chairman of the organization whose members included, in addition to Bush, such American policy leaders as Robert Patterson, William Donovan, future Supreme Court justice Arthur Goldberg, well-known psychiatrist William C. Menninger, journalist Edward R. Murrow, and movie mogul Samuel Goldwyn. The all-male, all-white group was formed in order to support Truman's remilitarization plans contained within a seminal national-security policy document, NSC 68. The present danger to which the group's title referred was "the aggressive designs of the Soviet Union." In his talk, Bush advocated peace through strength. To underscore the importance of citizen engagement, he pointedly declared he'd never been a "joiner," yet the magnitude of "the danger" of "aggressive dictators" around the world compelled him to break with his past practice. While he opposes the proliferation of nuclear weapons to other nations, he does contemplate using them to deter Soviet aggression around the world. Such was the logic of peace through strength that pursuing peace for Bush meant, "without question," avoiding war of the worst kind.

THE ATOMIC BOMB AND THE DEFENSE OF THE FREE WORLD

My Fellow Citizens:

The Committee on the Present Danger is beginning tonight a series of weekly broadcasts on the peril that faces the American people and how it can be met. The Committee is a non-partisan group of citizens who have organized to work together as the nation prepares to safeguard freedom. I have never been a 'joiner,' as we call them on Cape Cod, but I believe so strongly in what this Committee stands for that I am glad to be a member of it. We believe the nation's preparation to meet the danger must be on the same scale as the danger itself. We believe balanced armed forces are the heart of such preparation. And we believe the utmost speed is

---

"The Atomic Bomb and the Defense of the Free World," an address to the nation over the Mutual Broadcasting System radio network, on March 4, 1951 (Pamphlet by the Committee on the Present Danger, in Vannevar Bush files, Carnegie Institution Archives).

essential. I have been asked to begin this series with a reckoning of the probabilities of the defense of the free world, and how the atomic bomb affects them.

There is no doubt of the desire of the American people—and of our friends. We wish to avoid war. We wish to preserve our freedom and the free way of life. In a world where aggressive dictators are still at large, there is but one way to achieve these ends. That way is—to be strong. I am confident that the American people realize this. But we need to study just how to build that needed strength.

The key to the matter, in my opinion, is the A-bomb. At the end of the war our allies were exhausted. We disarmed. We know what has happened. Russia moved in. Working by intrigue and by the subversive overthrow of governments, she took over enormous territory and millions of people. But Russia stopped. Russia stopped at the boundary where the Kremlin was sure there would be war with us if it proceeded further. We saw the matter tested out at the time of the air lift in Berlin, and we know when we confronted the Russians with true strength, they did not force the issue.

The deterrent is nearly as powerful today as it was then. If Russia sent its armies rolling across the German plains tomorrow, we with our A-Bombs and the planes to carry them would destroy Russia. We could do it without question as matters stand today. We could destroy not only the key centers from which her armies would be supplied, but also political centers and the communications of the armies on the march. Initially equipped with weapons and supplies, those armies might keep rolling for a time, but there would be no Russia behind them as we know it today. The answer to this is that the armies will not roll. No all-out war is in sight for the immediate future unless they or we make some very serious error indeed. If Russia knows that she cannot go beyond certain boundaries without provoking a war, she will not pass those bounds; no war will occur. This has been well shown in recent years. The only apparent exception is in Korea, and there we did not make our position clear.

The difficulty is that we cannot count indefinitely on strategic bombing as the sole means of averting war. Today, it gives us a military stalemate. To maintain that stalemate is the real problem.

Defenses against strategic bombing have been mounting ever since the war. Jet pursuit ships controlled by ground radar can be enormously effective in bringing down high-flying bombers. Russia with its vast distances can have extensive early warning radar networks to alert its defenses. She can have great fleets of jet pursuit ships for defense, accurately controlled from the ground night and day. She can also have about her key positions modern antiaircraft artillery and also perhaps ground-to-air missiles. Russia in time can thus protect her key points. Note that I say in time. She cannot do it now. She cannot at any time safeguard all the places in Russia we might wish to attack. But in time there is a strong probability that she can defend key points to the extent that we could not penetrate to them without prohibitive attrition. She is also building a stock of A-bombs of her own. The deterrent of our A-bombs is real. But we cannot count on its remaining fully effective forever. I trust we have time—time to prepare the defenses that will continue the balance and avert war. But we do not have time to waste.

These defenses center in an allied army in Europe capable of holding a defensive line, stopping the Russian hordes if they should ever start, and so dissuading them from starting. That army must be well trained and it must be supplied with the very best weapons of every sort. It must be created before our present enormous atomic advantage is seriously lessened. Of course it needs to be combined with continued development of our striking air force and support for our Navy to keep the seas open, but there must be an army in being and on the spot capable of holding back the hordes of Russia. Such an army does not now exist.

This need by no means be a matter of opposing hordes by hordes. We have no idea whatever of invading Russia by land, nor should we develop any such idea. Ours should be a defensive line, a line to hold back the hordes while we strike by other means.

Many elements enter into this. In the first place, take the matter of tanks. Russia has 40,000 tanks of various sorts. All her military doctrine revolves about the use of tanks and artillery. But there have been developed in the last few years anti-tank weapons of great power. Relatively small recoilless anti-tank guns mounted on a jeep or handled by four men can put a heavy tank out of business, with a high probability of doing it before the enemy can get off his first shot, even at the ranges of 1500 to 2000 yards. These guns can be built in quantity by the hundreds for the cost of a single heavy tank. When the countryside is infested stingers of this sort, no tanks are going to roam that countryside long. There will still be a function for the light tank and for armored vehicles of various sorts. But the big tank has met its match, and unless techniques change in a way that I do not now see, it will become a liability rather than an asset in due time. I do not say that the big tank is now obsolete; I do say we can make it obsolete if we put our minds to the job and build the things to counter it. With that problem settled, the defense of Europe is simpler.

There is another factor, moreover, which is of enormously great importance. Out in the Nevada desert there have recently been a number of explosions. Presumably they mean the testing of new types of A-bombs developed by the Atomic Energy Commissions during the past five years. I will not speculate as to their nature, but we can certainly assume that we have not been idle and that we have more effective bombs today than we had five years ago. They may indeed be far more adaptable for a very important purpose.

We have thought of the A-bomb as a means for attacking great military production facilities or centers of political power. The A-bomb can also have important tactical uses. Suppose that a war were to break out three or five years from now and that the Russian hordes were held up by a much smaller number of well-disciplined and well-armed divisions. If the line were not too thinly held, if it were defended in depth with the land mines, anti-tank obstacles, artillery and other weapons that we can have if we choose, how would the Russians break it? They could do so only by a huge concentration of armies, artillery and tanks—the kind of thing the Nazis did in 1944 just before the Battle of the Bulge. But with A-bombs in existence this becomes a very different matter. An A-bomb delivered upon such a concentration by an airplane, or possibly by use of a gun or guided missile, would be devastating. In its presence, concentration of this sort would not make sense. Tactical use of the A-bomb thus will help to make the defense of Europe—with reasonable numbers of men—a practical matter.

Further, Europe is regaining courage and spirit. The mission of General Eisenhower and the evident determination of the American people are aiding greatly in that regard. We can join our strong and well-trained troops with those of our allies in Europe—we must assemble them in such numbers that they can hold the line. When enough men are mustered, there are important technical innovations to enable them to hold such a line against vastly superior numbers. It is not a matter of meeting hordes with hordes. Yet with even the most subtle of modern weapons there must be men to maintain the line and men to wield the weapons if they are to be effective.

I trust therefore that in our provision of manpower we will look well to the future. There is no thought in my mind that the men we bring in for training now will have to fight soon. Rather I think they will be the beginning of a well-trained, well-organized reserve. We need not only an army in being on the spot, but also behind it masses of trained men who can be called, if it becomes necessary, without a long period of indoctrination. If total war ever comes again it will break suddenly. I believe that the way to accomplish this purpose is to induct 18 year olds to have approximately two-years training and service and thereafter to go into the reserve to build up the essential body of trained men.

We cannot build the forces we need without sacrifice. This sacrifice must come in many ways, in foregoing some of the pleasures we like to enjoy, in increased taxes and heavier burdens, and above in the selflessness of our youth as they devote a part of their lives to training for the defense of decency and freedom. In my opinion we shall produce less interruption in the life of the youngster if we train him in the years of 18 and 19, after he has finished high school and before he launches his permanent career. Moreover the earlier the training starts, the longer will men be available for the reserve. It takes young men to fight a war.

As I said when I started this talk, the object of the free world is not to fight a war but to avoid the necessity of fighting. If we are wise I feel sure that we can avoid that necessity. We have today an able group of military leaders. We have a strength which Russia fears. The Kremlin will not strike unless it makes a mistake or unless we by the utmost foolishness cause it to make a false move in the belief that it can do so without bringing out retaliation upon it. We must keep such strength that we cannot be overwhelmed, such strength that to attack us would be suicidal. If we do, the attempt will not be made and we can live without a World War III. The sacrifices we shall make to that end, heavy though they may be, will be small indeed compared to the sacrifices we would make if through weakness or hesitancy we allowed a war to come upon us.

Nor does America stand up to Russia alone. Russia today faces the entire free world, of enormous production capacity and enormous numbers of men, a vigorous free world. The free world has no idea of making war on Russia. But it is determined to live in peace and to be strong in order to do so. We of the United States have great allies. They are temporarily in some distress for they suffered grievously during the war.

But their might is rising. France is rising with all of its great traditions of strength and independence. Britain is rising with its pride at having maintained the peace of Europe by its strength for many years. So are other allies as well. Their growing strength combined with ours can be made ample to stand off the present danger. If it is held in check that danger will in time fade. We will face the threat shoulder to shoulder, and facing it thus we will keep the peace.

32

A FEW QUICK (1951)

After Bush left official positions in the Department of Defense, he stayed in touch with military strategists, sitting generals, and those supervising innovations in new weapons and countermeasures. This long essay, which Bush privately and widely circulated among senior military officers and government officials, reflects to a degree Bush's own frustration— shared by other veteran leaders from World War II—over the poorer-than-expected performance of the American military in the Korean War. Of special concern for Bush (and others) was the battlefield stalemate that arose despite the superiority of U.S. weapons compared to those of North Korea and China. In "A Few Quick," Bush seeks to diagnose and treat the ailments afflicting the nation's armed forces in their Asian land war, arguing for the rapid introduction of prototypes and beta versions of new weapons designed for the Korean conflict. In his advocacy of bringing novel weapons quickly into action, Bush anticipated a popular practice of civilian technology leaders releasing beta versions to customers in order to stimulate a beneficial feedback loop. For the Googles, the Amazons, and Microsofts of the digital world, the pace of innovation can be so rapid that only by introducing partly finished products and processes can innovators fully take advantage of their advances, because creators learn from the experiences of "early adopters" or "power users." In the case of the Korean war, Bush argued that scientists and engineers could profitably conceive of the battlefield itself as a kind of laboratory, or "test bed;" and so by introducing selectively new unfinished weapons into the battlefield, under careful controls, military leaders would learn more quickly what new things work and what don't.

"A Few Quick" also addresses a challenge that bedevils military innovators to this day: the problem of working, however earnestly and effectively, on the wrong problem. Bush saw in the Korean War experience many examples of a mismatch between military innovation and the requirements for victory on the battlefield. He insisted that scientists and soldiers needed to get closer to the lived experience of battle and to the conditions of warfare in Asia. He also raised doubts about the value of relying so heavily on nuclear weapons when, as Korea showed, those weapons were not available for political reasons. So in "A Few Quick" Bush has two enduring questions in mind. The first is the perennial challenge for scientists and soldiers to identify accurately the conditions of battle and the technologies that can decisively shift the outcome of battle. Accomplishing this goal is easier said than done. The second challenge is the ever-present barriers—political, social, and cultural—to using nuclear weapons and the widespread view that such weapons are not usable in combat. Hence, Bush draws the conclusion that conventional military innovation deserves much higher funding and much greater attention than it normally receives. His view would find many supporters even now.

Finally, this is the first full publication of "A Few Quick." In his call for a revolution in technological innovation for national defense, Bush anticipates the creation of what is today known as the Defense Advanced Research Projects Agency, which is charged with conceiving, designing and building prototypes of breakthrough innovations that can alter the balance of power in combat. Radical new approaches, Bush insists, can be viewed by supporters of the status quo in weaponry as a dangerous form of dissent. "If this be heresy," Bush writes, "let's make the most of it." Observing that innovators can take pride in ambitious undertakings, even in failure, Bush notes, "Development, of a widely progressive sort, inherently involves mistakes and failures. So does research."

A FEW QUICK

We cannot stop Russia by mere masses of men. Neither can we stop her by mere masses of weapons. Our weapons must be better, far better. We

---

"A Few Quick," November 5, 1951 (Vannevar Bush Papers, Library of Congress, Box 137).

can have these better weapons in quantity if we really use our unique resources to full advantage. But we are not doing so yet.

What better weapons can mean was recently shown in Korea. After delay, after too much delay, we finally put in the field powerful combinations of modern weapons particularly adapted to the conditions there— unopposed air spotting, radar control, and artillery firing proximity fused shells. The ratio of Red causalities to ours went up fifty to one, and the Reds asked us to discuss an armistice. If we are really well armed the Reds will not force a world war on us. But we have got to get ahead, and stay ahead, in quality, ingenuity, and striking power of our weapons. This means more than A-bombs, important as these are. It means superiority in every phase of fighting.

We are now building the equipment for large armed forces, for ourselves and our allies, and we are faced with the old quandary: do we get masses of mediocre weapons, or do we get really advanced equipment which can give us the edge over potential enemies?

This is no small or simple problem. There are two extremes we might pursue, and neither one will give us what we wish. First we can construct a juggernaut of mass production, grinding out thousands of conventional units, carefully protected against those interruptions and alterations which slow it down severely, with the innovators carefully segregated in the background and allowed to play with their toys as long as they do not interfere with serious business. That can give us plenty of the wrong things. At the other extreme we could operate a production effort which is thrown off the rails every Monday by a Major with an idea, producing finally an aggregate in the form of a sort of Christmas tree to which the whole family have added gadgets in accordance with their own inclinations, a plane so loaded down with diverse junk that it cannot fly, a ship so covered with impedimenta that there is no room for crew. Neither of these extremes makes any sense whatever. Moreover the compromises we have employed in the past do not make much sense either. We need two kinds of programs simultaneously, two streams of effort, and they should be kept religiously separate. One should produce masses of the best equipment that can be frozen at the moment, and this should be protected against irresponsible interference. The other should produce a few quick. These few should be utterly modern innovations, to be subjected to test and scrutiny, and trial in maneuvers, as a means of arriving at the really new valuable weapons that will give us a striking advantage. There are plenty who understand the first stream well, and few who understand the second stream at all. Unless the second stream is vigorously introduced and supported, the first stream will take the stage, occupy all the industrial facilities, the money, the controlling organization, and push the second off into the wings.

This is a matter of vital national concern, not only because it goes to the heart of the quality of the new weapons with which our boys may have to fight, and the adequacy of the supply of these things, but also because it will affect the matter of costs severely. The taxpayer is a long way off when it comes to controlling costs, yet he pays the bill. Money from Congress is treated much more lightly than money that is saved out of a hard-earned profit and plowed back into a business. No governmental or bureaucratic control of production, no business monopoly in the production of goods for that matter, has the incentive or organization for holding down costs, or boosting production, that is found in good healthy competitive industry, as we have forcibly demonstrated to the rest of the world by our standards of living. Yet the production of munitions in great quantities must be government controlled and we have to make the best of it. Our costs and taxes will be high as a result. We merely insist that they not be astronomical. Our interests are much more closely involved than this, however. If we waste our defense dollars we can force the country into an economic tailspin, and that would delight the Kremlin. It is in fact exactly what they are waiting for.

Let us take an example. There is in existence a ground-to-air guided missile which proves to be a very important element in defense against high flying bombers. It is a powerful device capable of knocking a bomber out of the air with high probability of success at relatively long range. Even these, if fully successful, can by no means produce 100 percent protection, even for key cities, and even against high altitude bombers, and none against the lone bomber that comes in just above the land or water. Still they form one very important element in a complex of weapons which can give a significant defense to the key points we must defend against enemy bombers carrying A-bombs. We need them, of the finest possible performance within reasonable complexity, in quantity, soon, and at low cost. Why low cost, when the value of the areas protected is in the billions? Because the cost will determine how many we can have. It may well make a difference between inadequate and reasonably satisfactory protection whether they cost $50,000 or $10,000 a piece. And the way in which we go about producing them may indeed make all of that difference.

We know how to organize a stream of production to do this, reasonably well, and even under government control. It is adapted to the whole industrial system at which we excel. We know how to work such a system, to make it produce great quantities, to keep down costs in the process and even how to modify it deliberately for the relatively gradual changes that appear in evolving commercial products. But this guided missile is ready for procurement in the first stream of production. It does not need much 'few quick' treatment, for it is nearly past that stage. But there are plenty of other potentially valuable gadgets which do. We know much less about the second stream, and very little about how to adapt it to the bizarre needs of radically different military devices. We need to produce anti-aircraft guided missiles in quantity, and cheaply. At the same time we need a 'few quick' of better missiles for the same and other purposes.

There are a few principles we can write down for guidance. The two streams of production must be independent, and their interaction must be very wisely controlled indeed. The status of the second stream must be equal to that of the first. It must have funds, organization, equipment. Above all it must have independence. If it is secondary to, or controlled by, those who operate the first stream it will not long exist. The two streams should meet only at the very top echelon, where their interaction at every level should be controlled. Moreover this top echelon must be manned by those who thoroughly understand both streams. Here is the rub. There are very few who understand both streams.

An example or two from the last war will show the need for independence. There was then constructed a 'few quick' organization, to aid in the transition from the laboratory to use. It was built as an adjunct to the Office of Scientific Research and Development, which was itself independent of both the military organization and the whole gamut of civilian control organizations centering in WPB [War Production Board]. It was a small effort, this 'few quick' affair, with little money or facilities, slipped in more or less furtively as a part of development, which conveniently proved to be a rather broad term, for OSRD was not supposed to produce. The great organizational machinery for procurement hardly noticed it, the public never heard of it. It had little of the protection on materials or personnel accorded to the regular channels; in fact it had to battle through the maze of priorities for its materials and through Selective Service to keep from losing its young personnel to bear guns, so that we might have another battalion in the line. It was not manned by the great masters of American production; they were busy elsewhere. It was not formed by the military organization or by the WPB; in fact, they hardly recognized its existence; it was too small. It was formed in fact by a group of young physicists and engineers who were developing short-wave radar and who recognized they needed a shortcut around the ponderous regular organizational machinery. In its small way it produced results.

Ground Controlled Approach, GCA for short, that radar system which can talk down a plane to a landing in fog or murk, now used in all principal airfields, appeared in the laboratories during the war. It excited no real attention as an idea, or even when tried in breadboard form. Fighting men and producers were much too busy to explore queer ideas like this. So a 'few quick' were made. One of these sets in England brought in safely a dozen B17s that were lost in the fog and about to be ditched in the sea, whose pilots incidentally had never heard of the device until they were talked to and told about it and then talked down to a safe landing. Then everyone wanted it at once. The only sets in existence then or for some six months were a handful that had been made by 'few quick.'

Loran appeared in the laboratory, developed by British and American radar men, growing out of the Gee system to control bombers. It could give accurate position to a ship or a plane in a few minutes, a position within a few miles in the thousand mile reaches of the open ocean, in any weather. It could enable escorts to meet convoys, and long-range planes to proceed surely to the point of sighting enemy submarines. It could render the flights of air transports far less hazardous. Again it excited little enthusiasm at first; the Navy started a long drawn out set of comparisons with conventional methods of navigation, incidentally attributing all of the errors it found to the new device. The only set in operation was an experimental one, manned by laboratory men, with widely spaced stations covering a part of the North Atlantic. Then suddenly all resistance and inertia disappeared and everyone wanted it at once. There were only six sets of the main equipment in existence, all made by 'few quick,' and no more available until the wheels of production could grind. There was a conference, an animated one, to decide whether the meager equipment should be all used in the anti-submarine campaign, or in the Pacific or the facilitate the plane transfer from Brazil to Africa. I presided at that conference, and found it useful to remind the military men present that my agency owned the only sets in existence, and that agency reported only to the President of the United States. It had a very salutary effect.

There were other examples. 'Few Quick' not only got our models for initial use, it enormously speeded up the subsequent regular production. Production models appeared in months instead of a couple of years. In these subsequent moves, individual flexible industries helped enormously, large ones that had no hardening of the arteries, but notably small ones, light on their feet, used to preserving themselves in the maelstrom of American industry by their speed, versatility, courage and luck. But it never could have been done before if there had not been independence— and nerve to start the ball rolling, and set the example. This required young men with independence and full support.

The first stream, that of deliberate mass production, should be manned by men who know their business very thoroughly indeed, who have lived their lives in the intricacies of the American method of making very large numbers of devices cheaply and well. There should be no compromise on this point. Bright idea men, in uniform or out, should be allowed to observe and comment, but not to butt in. The control should definitely be in the hands of those who know the mass production game. The converse is equally important. The second stream, that of 'Few Quick,' of end runs, of experimental production, is very different in its entire concept. It is dangerous to man it with men who know too much about the subject. If this be heresy, let's make the most of it.

Under the National Defense Research Committee, a part of OSRD, the independent civilian organization devoted to the purpose of development of new weapons, there were 19 divisions. All of them accomplished valuable results. But it is notable that those who went furthest, those that accomplished rather startling things, were manned almost entirely by men who, when the war began, knew practically nothing whatever about the subject for which they later took great responsibilities. Of course, there were divisions on subjects where the opportunities for radical advance were not great, in fields that had already been well plowed. Of course, the greatest opportunities lay in the areas that were so new that no one whatever knew much about them at the outset. But on the record the conclusion is inevitable that, when one wants a really startling departure from the pedestrian process of improvement in a technical area, it should be delivered bag and baggage to a group of young keen men, hardworking and intense, thoroughly competent and sound in fundamental science that underlies all technical advance of the sort, hardheaded on practice and costs, uninhibited, full of ginger, balked by no traditions of what cannot be done, holding with scant respect the old masters of practice in the field with the absolute minimum of discipline and control consistent with having any organization at all. The group should of course be backed up very thoroughly indeed, protected against predators, protected against interests that know all the answers and resent the parvenu. Managing a group of this sort is not quieting to the nerves. Rare, but existent is the older man who can fit into such a group and provide perspective and wisdom without putting on a brake. Some of the results will be bizarre and useless. But the occasional product of such a group is worth all it costs in money or headaches. The trouble is that regular complex organizations very seldom are willing to accept the headaches.

Two examples of the free-wheeling end run can be cited from the last war. One gives the example of heartening collaboration, the other does not. The first one is the proximity fuse. The idea of such a fuse appeared in a number of places early in the war. It may be noted in passing that, once the stage is set, there are always plenty of ideas, especially half-baked ideas, on the subject of new weapons. The difficulty is to select sound ones, and above all to make them work. Still more difficult is it to get them into use fast; industry usually counts on five years at least from an idea in the laboratory to a product in sales. During the war this was compressed down to two years or even one and sometimes on rather simple things to less than that.

The idea of the proximity fuse was to put a radio set in a shell so that when it came near an object it would explode the shell. Anyone who has nursed a sick radio set in the early days knows that it could be made to squeal, and that its pitch would change if one moved a hand near it. A so-called musical instrument works this way, and does not sound badly at all. The proximity fuse was to include, to oversimplify a bit, a squealing radio set, and use the change in the squeal to fire the shell when it was just near enough to a plane to do the most damage.

The proposal was absurd on the face of it. Radio sets contain glass tubes, and delicate filaments. They were notoriously temperamental. It was proposed to compress such an affair down to the size of a tobacco can, put it in a shell where it would get a jolt when the gun went off as though it had been hit by a hammer, and then have it in precise and delicate adjustment after it was on its way. Moreover, it was proposed to do so safely, so that it would blow up the enemy and not its friends. It was a very wild idea if there ever was one. Nevertheless a group of wild, but very sane and sound, youngsters did just that.

But it is one thing to develop such a device, and quite another to put it into production and produce millions of workable units at a tolerable cost. The British started the development, and abandoned it. This was wise; in fact a number of things were left by agreement to one country or the other; for it never could have been produced in Britain under the conditions then prevailing. The Germans worked away at it, and did not get within measureable distance of success, fortunately for us. The story of the development and production of the proximity fuse in this country is a classic. Success was due to the finest sort of collaboration between the Navy, the young group that did the development, and a number of highly effective industrial organizations. By agreement, in fact on the insistence of the Navy, the whole thing was left under the control of the original group of scientists and engineers, right through to mass production. The Navy assigned first one, and then another, highly capable officer to work in the civilian organization and help keep the ways greased, men who have enviable records in other endeavors since. A banker, not so young in years but young in spirit, helped keep the relations on a cordial basis. But the whole affair revolved about a young physicist, surrounded by a group of young scientists and engineers who respected him enormously, but were careful not to show it too much. In fact there were two such groups, for there was a parallel effort with many of the same characteristics working on proximity fuses for bombs and rockets; but the story would be too long if we included them both.

There were obstacles of course and some of these might have been fatal if it had not been for the steadfast support of the Navy for an effort controlled by an entirely independent agency. Selective Service, in its then absurd form, nearly took out the very key men necessary for success. Local boards could hardly understand such things, in fact they could not be told the whole story for reasons of security, and the central organization was not shot through with vision or statesmanship. We haven't learned our lesson on this subject even yet. For a long time the restrictions on use of the device were cripplingly severe, especially on land, for fear the Germans or Japanese would get it and stop our bombing. Highly placed officers, with little background in science or production, would not accept assurances that neither enemy could make the things within a couple of years if given samples or blueprints. Omniscience in top circles is always a hazard. Tests, production controls, safety provisions gave difficulty. But the project was a success.

At sea it produced a factor of perhaps five in the effectiveness of 5-inch anti-aircraft guns. That is, it multiplied by five the number of birds brought down by a given number of rounds. It thus saved many a ship. It brought down, in combination with other devices, over 90 percent of the buzz bombs that were shot at Antwerp and came within its range. At the Battle of the Bulge it caught German troops in the open in the fog and decimated them, for it could work in proximity to the ground, and accurately placed air bursts are deadly. It probably saved Liege. General Patton, after watching this performance, remarked that the entire tactics of land war were due for review in the light of the greatest improvement in military artillery since the introduction of rifling, except that his language was more terse. It was one of the most successful devices of the entire war.

When it was introduced in Europe there were teams present to watch for German counter measures. None came. The Germans could not believe that the Americans had produced any such thing, and they argued about it until the end of the war.

The other device was equally successful, but is not as pleasant a subject. This is the story of the DUKW. This is an army truck converted into a boat, so that it can turn on water or on land. It is one of a family of vehicles, amphibious jeeps, weasels, etc., but the DUKW story best illustrates the points that are being considered. These vehicles were developed by a group of young engineers in NDRC, headed by a chief who was not so young, none of whom as far as I know had ever before developed a land vehicle of any sort, or had in fact been in any unit of our great automobile industry. They had lots of sense, and a perfectly astounding amount of drive and push. Some units of the vehicle industry collaborated with them well; some certainly did not.

When the DUKW was first proposed, a General, high in procurement, remarked with some emphasis that the Army did not want it and would not use it if it got it. A General in NDRC, under orders, voted against starting the project. Some men, in the automobile industry, snorted audibly about the wild group of amateurs on a rampage. Some of the young group undoubtedly did do very unconventional and annoying things, entirely legal of course, but viewed with disfavor by those who had more serious programs to further, and there were occasional demands that they be tamed or suppressed. Even some of the independent scientific groups, on their own bents, were not sympathetic to some of the manifestations of youthful exuberance. It was a long uphill push. There was one General in the Army, a man of vision, determination and authority, who saw the picture from the start. Except for him there would have been no DUKW.

The tide turned when there were some tests on Cape Cod, trying the unloading of a Liberty Ship across the beach. When the time of the test came a full gale was blowing. During the gale, late at night, a Coast Guard boat with seven men aboard went ashore in the surf on a bar. It was too rough to launch a boat from the shore, and they could not be reached from the sea. Mind you, at this time, it was not yet known how a DUKW would behave in surf. At 1 o'clock in the morning the young engineers took a DUKW to the scene, ran through the surf, picked off the men and brought them ashore. In the morning the craft was gone. Moreover, an enterprising photographer took pictures, for the group during development were not unmindful of the necessity for sales effort under the conditions they struggled with. A day or two later, at a Cabinet meeting, in the presence of the Secretary of the Navy, the Secretary of War showed the pictures to the President of the United States, and remarked that it was probably the first time in history that a Navy vessel had ever been succored by an Army truck. After that much of the obstruction disappeared. Not all, for a General still categorically refused to allow any of the young civilian developers to enter an active theater where crews were being trained to use the vehicles. They got there, but only by loaning them to the British, who in turn loaned them to the Americans. I have always wondered whether the General, whose name I forget, ever found it out. I should remark that he was a General rather fresh from civilian life, not a professional soldier. Before the show was over the young prime originator of the DUKW sat in staff conferences in five theaters.

Without the DUKW the Sicilian landings would have been in grave difficulties. After it was over, General Eisenhower asked General Marshall to congratulate the officers in Ordnance who had developed it. He picked the wrong group, but the right youngsters were commended. The Normandy landings took quite a different form because DUKWs could move cargo straight from the ships, through the surf, to the ammunition dumps and the troops. On many an island of the Pacific they introduced an essential element into modern amphibious warfare.

Yet the DUKW was no great shakes in the application of science to warfare. Anyone can make a truck body like a boat and add a propeller. Anyone who can work out a scheme for rapidly altering tire pressure to pull through surf and sand. Anyone can see that such a device, which does not lose traction when a wave hits it, can go through surf. It might take a little vision to see that it could go through surf ten feet high. It took more vision to see the need for such a device clearly, long before it could be used. And it took a versatile determined independent group to push such an idea from the drawing board to use through all the maze of official skepticism and preoccupation. Such a thing will be done in regular channels only rarely and when there is a very extraordinary individual present somewhere to see through it. Occasionally it is done in the Services, when a keen group of young officers escape sufficiently from control and risk their careers; witness the Navy's very early development of radar. But the smart way to get rapid advance in an unconventional way is to give a group of sound youngsters their heads. They must distinguish the really practical idea from the thousands of screwball proposals that always abound, but if they are a sound group they will.

It is far more difficult to accomplish this sort of thing in peace time than it is during a war. In fact it cannot be done the same way at all. This article is no plea for the organization of a new OSRD. The confusion of war produces a broken organizational field, where forward passes and all sorts of unconventional plays are possible. Moreover, during war men will take a risk. In days when lives are being risked abundantly individuals will risk their health or their time or their reputations. Patriotic interests will produce, from the most unexpected places, groups that will grab a ball and run with it. The air is so full of reverberation that the ridicule of the amateur effort is hardly audible. But it is different in the grim days of uneasy peace. How, then, do we get on with it? It can still be done, but not in spectacular manner, nor with the same flair, nor with the same sort of organization.

To do the job today it has to be done within the organization of the Department of Defense. There is no use baying at the moon. Congress will not set up in peacetime an independent organization with the necessary freedom and funds. Franklin Roosevelt is gone, with his flair for the original, and his backing without which an independent organization would be left on the sidelines. The bureaucratic controls of Washington would smother any small independent outfit doing queer things. We have to face reality.

Fortunately the case is far from hopeless. Throughout the Services there are strong keen groups of young officers with ideas. Some of them are now pushing new and important devices, not for their publicity value as a basis for spectacular stories, but because they will work. There is a path, clear to the top, through the Research and Development Board [of the Department of Defense]. Various teams of civilian scientists and engineers are in touch with progress, in advisory committees and the like. Here and there real progress is in fact being made. All it needs is a bit more of unusual organization to give groups of this sort their heads and a scope of effort commensurate with the opportunities before us. This means money for development, opportunity for rigorous testing in the field, freedom from the absurd notion that failure in research should wreck a career and above all access to the top through a line of authority that does not contain a single roadblock. It can readily be done, if those at the top really wish to do it, and if they understand it. Some of them do.

There is plenty of opportunity left. The art of war is still changing; the days of invention are by no means over. Conditions vary among the Services. The most spectacular opportunities probably lie in the Navy and the Air Force. But the opportunities in the Army are by no means negligible. Let us pay our attention principally to these, and to the places where Army and Air Force overlap, unfortunately with a connection which does not operate any too well.

Are we at the end of airborne operations when we drop men in parachutes and tow gliders? How about helicopters as exceedingly mobile artillery, mounting recoilless guns. Or planes that shed their wings and become vehicles. There has been much speculation on a plane that could rise and land as a helicopter, and twist its wings for rapid flight; what is it good for in military operations? We have ships for plastering a landing beach with rockets, how about a cargo plane that could similarly blast a road or concentration.

The last war produced quite a development in heavy cavalry, in the form of wide tank sweeps. How about light cavalry? Both forms had their uses and doctrine in the last century. What is the modern mechanical equivalent of light cavalry? Did we exhaust all our ingenuity on methods of getting over rough ground when tracked vehicles were introduced?

There are plenty of other ideas. I have no really startling ones, if I had I wouldn't put them in this article for export to Moscow. But turn a young group loose and ideas will abound. Some of them will be of great moment.

The old fields all need to be combed. Let us take the case of the tank. It is an excellent case, for the tank is bound to go through some sort of metamorphosis. Anti-tank guns and ammunition have appeared which alter the entire situation. When a relatively inexpensive gun can destroy a tank at 1500 yards, with a high probability of doing so before the tank can get into action, when that gun can be handled by four men or mounted on a jeep, one section of the art of war has altered enormously. This is by no means to say that the tank is obsolete. But when a jeep meets a tank costing 50 times as much, and the outcome of the engagement depends almost entirely on which gets in the first well aimed shot, and practically not at all upon whether the tank carries two inches of armor or ten, the whole subject of tank warfare needs prompt and rigorous review. Something else new may come in to alter the picture—one should never depend upon a single line of technical development—but as things now stand it certainly looks as though the heavy tank would be obsolete as soon as such guns and ammunition are widely available in great quantities. There is not much use in waddling about in enormously thick armor if there may be dozens of guns in the hills or among the trees than can penetrate it. This is all good news, for the Russians have many heavy tanks, and their whole doctrine seems to revolve about concentrations of men, tanks and artillery. We are certainly not through with the vehicle armored against machine guns and fragments, of high mobility, long endurance, and long range. But we need to take a hard look at the whole tank picture.

As a matter of fact the whole tank picture required review the moment the first bazooka appeared, for there were bound to be bigger and better bazookas, and that was eight years ago. When an enemy could rise from a ditch by the side of the road, and put a tank out of action with a shoulder weapon, it was obvious for one thing that a tank that could only look straight ahead, or could look to the sides only crudely, that had no really versatile anti-personnel weapons when buttoned up, was as obsolete as the dodo. We have been building much the same tanks ever since. Now range and precision have been added to ability to penetrate and we had better watch our step.

We are building many tanks of the old sort, at enormous expense. We are improving these at the slow pace at which we in this country introduce innovations in passenger automobiles. Well and good, we have to put into production the best we have at the moment, for tanks of all sorts are not obsolete for all uses. Well and good, provided we don't build too many, at a half million dollars a piece, and provided that is not all we do.

How about steam-driven tanks? I can hear the automotive engineers snort, but I'm not so sure. At least the fuel would not need to be volatile and explosive, and most tanks are lost by fire. Moreover, I remember the steam automobile as quite a gadget, and I have a feeling that, if it had been given the same concentrated attention in development, and not barred by trick laws, it would be a better automobile today than the ones we have. At least it wouldn't require fancy gadgets, in the form of "flowomatic" drives, to make a constant speed engine handle a variable speed load. I'd like to see a young group tackle it, if they were really free from the frowns of those who know the steam engine is dead. Or, if we want to be more moderate, we might give the gas turbine a whirl.

But the tank affair needs more than end run developments. In the first place there needs to be an analytical review of the whole affair from the standpoint of military tactics, backed up by experimental maneuvers and tests, backed up whenever possible by actual use in the field if, in the next few years, we are forced into action like that in Korea. There will be no single solution, for conditions of campaigns, the nature of terrains differ widely. This is primarily a military job, but civilians of background can participate to advantage. In fact unless civilians participate, there is danger of decision by weight of stars rather than weight of evidence. The military men should come from field experience, from military schools, from ordnance laboratories. Above all, they should be the keenest group that can be assembled. They should argue interminably in the evening. They should be passionately devoted to a wide range of theories and doctrines. There are plenty of keen officers of this type, but they don't always get the chance to operate without the heavy hand of reactionary authority upon them. In fact, in military circles, the keen youngster cannot be nearly as aggressive as he can in a scientific laboratory, for it is only too easy to sidetrack the unconventional individual, and youthful intellect is sometimes a trial to authority. However, there are youthful and intellectually courageous officers, of every calendar age, in the Services. They can do the job without question or peradventure, if they are ordered to do so by the top echelons and given the necessary freedom to work. This is the first part of the job.

Next, we should go ahead with production of types of tanks that are undoubtedly still useful. This is not difficult, for in the vast range of character of military operations we may be engaged in, several types are undoubtedly in this class. Also we start almost from scratch—we have very few modern tanks available for combat—so we can build more with a clear conscience. This schedule of production should be accompanied by a deliberate program of orderly development and improvement. Innovations and changes should be introduced at calculated intervals, to obtain the optimum end results, neither freezing for undue lengths of time, nor breaking the production lines in a continual turmoil. As this procurement proceeds emphasis should be shifted from time to time, among size and types, in accordance with the results of parallel analysis and experience. This program should be in the minds of old-time production men, from top to bottom. Cocky individuals—that know all the answers—should be securely excluded. A separate group from the regular line, reporting to the top, should be continually analyzing performance and costs, and especially costs, and then conclusions should modify the introduction of changes, and the emphasis as between contractors. This is what we would ordinarily do, and reasonably well. It is not enough.

In addition, and simultaneously, there should be independent groups, working out designs of radically different tanks, of radically different fighting vehicles of every sort, for the mobile gun merges gradually into the tank. There should be several of these groups. They should be quite independent of the regular production program and controlled so near the top that they are not under anyone whose job it is to procure tanks. Fortunately we have the nucleus of such groups, and they are already moving ahead along these lines in spite of the obstacles of being more or less buried. Fortunately also there are officers of vision and understanding, to man the lines of special control if they are ever created. These exploratory groups should not be expected to produce anything useful for a considerable period. There should be no penalty on their failure to produce anything at all, if they spend government money economically and within the rules. Above all these groups should not be part of the regular automotive industry of the country. Anyone who knows all about vehicles should be securely barred from control of their affairs. There are not many industrial organizations in the country that will stick their necks out to the extent of taking a contract along these lines, but there are some, and if they do not exist they can be assembled for the purpose. They must have on their staffs highly competent applied scientists and engineers, metallurgists, specialists on plastics, mathematicians—yes mathematicians, chemists, thermodynamists, specialists on the physics of the solid state, crystallographers, a dozen other queer ducks, either full time or partly available as consultants. They should be assured of the best advice possible from automotive engineers of every stamp and of the highest caliber—but advice only. They should not be committees; these produce compromises, not radical progress. They should have experimental production functions and test facilities. They should not be given a set of "military requirements" of the usual sort. More initiative has been balked by arbitrary military requirements than by any other device invented by the mind of man. However they should be given access to the whole range of analytical studies in their formative stage, invited to present ideas and proposals, and incidentally paid enough to cover their costs while they study and prepare to do so. When an attractive plan appears in this manner they should be given a contract to go ahead.

We can lean on our allies for some of the unconventional effort. There is no group better at innovation than the French, if they are given the chance. The Italians have always excelled at unusual automotive vehicles. Certainly after our experience as partners in the war, we pay tribute to British initiative and common sense. In fact many of the most important advances were carried out with such close collaboration that it was impossible after the event to assign priority of invention to either British or Americans. One of the most pleasant experiences of the postwar world has been the almost complete lack of national claims in connection with the joint accomplishment. Certainly with our encouragement and welcome there can be innovations produced, in connection with our military aid program, which are valuable indeed. But if they have to pass the gauntlet of appraisal by some subcommittee of some Joint Staff, they will not get beyond the idea stage.

What will these unconventional efforts produce? I have no idea. Perhaps they will not make their armor out of metal. Maybe they will make a vehicle that will run on treads one side up, and roll over on its back to proceed on wheels. I have no brilliant ideas. But if there is something radically different in the form of armed vehicles, a wide departure from conventional practices, they will find it and shake it down. Then at the proper time, if it really works in the field, it can be inserted in the regular stream of procurement without a jolt. Their main contributions may be far from the subject of vehicles, far from any of the lines of development we now regard as established. Perhaps they will find nothing. In that case, we will have some assurance that we are not overlooking a bet which a potential enemy might stumble upon.

This is the sort of thing at which we can excel, but we have to organize intelligently to do it. No totalitarian state can perform that way. In a totalitarian state, they give a man a job to do and if he doesn't perform satisfactorily in the judgement of some commissioner they shoot him or take away his food cards. No totalitarian state can organize to provide for expected mistakes and failures in operation, it is much too subtle for a police state, there is too much jealousy and distrust about. Development, of a widely progressive sort, inherently involves mistakes and failures. So does research, but in research they can often be covered up by grandiose claims and arguments. It is hard to bury a useless and bizarre gadget before it is seen. Here is an area where we can compete with every advantage on our side. But to insert it into the dignified methodical process of military procurement will take courage and vision. Fortunately, to try it will not cost much—in comparison with overall costs of preparing to defend ourselves. After all, OSRD in its whole history spent only half a billion dollars, and we talk [today] of military budgets of fifty billion. There is a start already in the Services. It needs to be recognized, expanded, brought out into the light, and given the freedom which is essential to its performance.

33

ON LEADERSHIP AND MANAGEMENT (1951)

In this letter to the chief of research at Merck, then as now a leader in pharmaceuticals, Bush shares his own refined ideas on how to succeed as a research manager. Bush's advice was valued, not only because of his own success leading research during the war, but also because he served on Merck's board of directors and was close to the company's namesake, George W. Merck. Here Bush draws partly on lessons from his wartime leadership work and his university experiences as a research leader. He provides pithy maxims such as, "No decision should ever go above the level at which there is a competent man to settle it." He also displays a sense of humor about the burdens carried by leaders, writing, "When he is away no one should see the need to call him up unless the place burns down or the president resigns." Despite the passage of time, Bush's advice seems fresh and relevant to today's managers of research and development.

ON LEADERSHIP AND MANAGEMENT

When we last conferred, I looked at our organization chart and could find nothing wrong with it. There is an old saying, however, that the danger of

Vannevar Bush to Randolph Major, December 29, 1951 (Library of Congress, Vannevar Bush Papers, Box 69/1686). Bush served as chair of Merck's board from 1957 to 1962, following the death of George W. Merck in November 1957.

an organization chart is that it may be mistaken for an effective organization. I hope you won't mind if I tell you frankly that I fear we may have such a situation here.

I have several general impressions. One is that you have been enormously valuable to the company in the past, that no small part of this has been due to your standing in your profession and your contacts with scientific men, and that the burdens of your present duties now interfere with the most valuable part of this activity. Another way of saying much the same thing is as follows. In the past, the company has showed great vision and resourcefulness in the selection of vigorous pursuit of its objectives, and this has been to a very considerable extent due to your grasp of scientific progress everywhere, and of the opportunities it opened up. The company faces today and faces acutely the same problem of working out its program for the future in the most intelligent possible way. Yet I do not find your influence in the top councils on this subject nearly as complete or effective as I feel it could be, and I believe this is because you have allowed yourself to be buried in a mass of detail to the exclusion of more important policy considerations. I suspect you are ferreting into small matters when you ought to have your feet up on your desk thinking, or arguing with some of the best chemists and medical men in the country who are your friends.

I believe the basic reason for this situation is that the size of your operations has grown, while your management of it, operation charts to the contrary notwithstanding, is much the same as when it was small. Thus you may be applying to a complex affair methods which work well in a simple case, but which break down when many men are involved.

Thus far I have written only of impressions and generalities, and touched on some of the same points when we last met. So let me be more specific as prelude to our next discussion in which we can delve somewhat deeper. Perhaps the best way will be for me to state a few simple principles of management, see whether you agree with these, and make some queries as to how they become applied. I'm sure there is something very important here for us to discuss.

The first principle of management of a complex affair is to have good men about. Here I feel we are on reasonably solid ground. The men under you are a varied group, but as far as I know they are able in diverse ways. I could criticize, but I think the difficulty does not lie here. In general it seems to me you have men under you of ability and characteristics ample for your purposes. Some of them seem to me to be decidedly outstanding. Since you have yourself pretty largely selected and trained this group, I judge you agree with me on this point.

The next principle is that one must delegate effectively. One part of the art of delegation is that it must be clear cut. Every man must know exactly his duties and responsibilities, must have precisely defined authority, must know to whom he is responsible, and his correct relation with those on his own level. Is this really now the case? Could we pick any one of your subordinates and find that he knows exactly where he stands, what he is supposed to accomplish and how? I have studied our organization manual, but the point I have in mind goes beyond this.

When one delegates he must back up. When one places in the hands of a subordinate the authority to make a decision, that decision should be reversed only very rarely and under the most extraordinary conditions. When one leaves to a subordinate the decision on a matter and the decision has been made by him and become known that subordinate's judgement must be taken and used even if one thinks it is wrong and that a mistake has been made, and the decision should be reversed only when it is genuinely essential to do so in order to avert catastrophe. The guidance of subordinates should be by teaching them to know and follow the chief's policies and thoughts, not by interfering with their operations. Now I cannot testify on this because I do not know well enough your methods of dealing with your group. I do know that, unless this policy is adhered to strongly, the men under you will hesitate to take decisions at all without checking with you in advance. I have seen situations where this occurred and the result was always chaotic.

Throughout an organization, the contacts of the chief for the purposes of information should be uninhibited, while business continues to be handled in channels. In particular every man in the organization should feel he can take a personal problem to the chief and be received sympathetically and without it backfiring on him. This is possible only if lines of authority are so clear cut that no one will misunderstand when the chief wanders about, for he never gives orders or butts in during such explorations. Will you give a bit of thought to your own direct relations and tell me how this actually works out?

Units in an organization should be made as autonomous as the capabilities of those in charge of them will allow. No decision should ever go above the level at which there is a competent man to settle it. Units should handle all their internal affairs, and be responsible only for general results and progress. How completely is this now the situation? The chief should not sit in where a subordinate unit is doing business. Otherwise it won't take the full responsibility it should. Moreover competent men do not like to be watched in detail by their chief as they proceed, for it tends to undermine their own control of their groups. I made a couple of first-class errors in this regard myself during the war, and they certainly took me into trouble and taught me plenty.

Line and staff functions must be separate and clear cut. A staff officer speaks for his chief when a decision must be made which is within the purview of no one subordinate division. He cannot also be in the line organization. Oftentimes a one-man staff is sufficient. But he must have the confidence of his chief absolutely. I can't make out how you have the staff function provided for.

The chief's office should be simple. It should contain no function that can be placed elsewhere. Its function is to aid the chief to do his job, mainly by freeing him so that he can concentrate on important matters. Just because a function occurs in several divisions is no reason for putting it directly under the chief. For purposes of business the chief should not have more than seven, preferably five, reporting to him. For advice and consultation he should spread broadly. He should not be bothered with a lot of clerical or auxiliary operations in his own office.

The chief should have leisure. Half the contacts he makes in a day should be on his own initiative, and as he thinks through to a conclusion a really important policy or program. In pursuit of such an objective he should be able to absent himself for days on occasion, without prior arrangement or special provision, for his organization should be such that it will run itself as far as all ordinary matters are concerned. When he is away no one should see the need to call him up unless the place burns down or the president resigns. The chief's organization should be used by him to accomplish worthwhile things. It should not use him, or weary him, or be a constant nag.

I could go on. But I think we have enough here for a start. Won't you look these over and do a bit of introspection so that we can discuss these points and others frankly. I trust you realize that my entire project is to aid you if I can.

34

"THE TIMING OF THE THERMONUCLEAR TEST" (1952)

In 1952, Bush led a secret campaign to delay, postpone, and perhaps permanently halt the first test of a hydrogen explosive device. This memo, which was principally written by Bush, presents a forceful argument that any test of a hydrogen explosive in the final weeks and months of Harry Truman's presidency would represent a significant missed chance to slow or halt the nuclear arms race. In private letters and conversations, Bush argued that any test would lead to a "point of no return" after which an escalating race for more and more destructive H-bombs would be inescapable. The historian Barton Bernstein, in a paper on the effort to prevent President Truman from testing an H-bomb, concluded that "the idea [to do so] came first from" Bush, who in turn enlisted J. Robert Oppenheimer, chief scientist on the Manhattan Project, in the secret campaign. Oppenheimer and Bush, who had approved of the use of A-bombs against the Japanese, persuaded Truman's secretary of state Dean Acheson to create a "panel of consultants," chaired by Oppenheimer and including Bush, to deliver a recommendation on whether or not to test. In September 1952 (most probably), the panel argued for delay in writing, forcefully insisting that, at the very least, the incoming administration, which was all but certain to be led by Dwight D. Eisenhower, should make the test decision. Bush's argument was widely accepted. The chief aide to U.S. senator Brien McMahon, who chaired the congressional Joint Committee on Atomic Energy, summarized Bush's plan in May 1952: "There's a move afoot," wrote William L. Borden to Senator McMahon, "to avoid conducting any major H-Bomb test—on the theory that the Russians are never able to do anything until after we have done it and therefore if we never test the H-Bomb the Russians never will." That was Bush's position in a nutshell, and the implications were sweeping: with a halt on the test and refinement of H-bombs, the United States and the Soviet Union could more easily manage atomic weapons and perhaps more effectively deter other nations from building them.

Alas, President Truman rejected the arguments of Bush and his fellow dissenters, who, in addition to Oppenheimer, included Allen Dulles, a CIA official and brother of John Foster Dulles, who Eisenhower would select as secretary of state. Rejecting Bush's argument in favor of delay, the president gave final approval to the first H-bomb test, which took place on a Pacific island on Oct 31, only days before Eisenhower, the Republican candidate, was elected President. Eisenhower briefly considered a halt to further H-bomb testing, but he instead supported a vast expansion of H-bombs, and sanctioned a massive number of above-ground H-bomb tests, which led to both enormous environmental damage and worldwide protests. Moreover, as Bush had warned might occur, the Soviets reacted to the American test by increasing their own research and development on an H-bomb. What follows are excerpts from the secret memo, first declassified in the 1980s. Reflecting Bush's core arguments, and relying on much of his language, "The Timing of the Thermonuclear Test" is a significant document in the history of nuclear weapons and, more broadly, in the history of humanity's struggle with new destabilizing technologies. Should a new technology be embraced merely because it exists, or can the technology be rejected or at least tightly controlled? The question continues to challenge humans to this day. Emerging techniques of gene editing and bioweapons present comparable questions, and especially in the context of continuing doubts and fears that bans or halts on emerging technologies are very difficult or even impossible to enforce. The core question continues to be, if we can build a new thing, must we build this new thing, and must then we use this new thing? In the case of the H-bomb, the answer was yes.

Finally, "The Timing of the Thermonuclear Test" is important because the document conveys Bush's trademark clarity of analysis, and his sober handing of complex issues arising from the paradox that great science and engineering can lead, perversely, to unimaginable destruction.

"THE TIMING OF THE THERMONUCLEAR TEST"

THE PLAN FOR A THERMONUCLEAR TEST
CALLS FOR A NEW DECISION

Character of the Test: A test of a thermonuclear device is planned by the U.S. government for the month of November 1952. This device is the product of many years of study, culminating in two and a half years of intensive technical effort which began after the Government's decision in 1950 to proceed with the development of a thermonuclear weapon. Great technical advances have been made in this period through a combination of good luck, great skill and high dedication. This first test may not work, but among leading students of the problem there is now very little doubt that the scientists concerned are on the right track.

If this test is successful, it will have the explosive power of 100 to 1,000 times as great as that of the atomic bomb used at Hiroshima. It will thus be something more than one more in a series of scientific tests. It will be impossible to conceal the fact that this event has taken place, and very difficult to conceal the fact that it is an event of great portent for all men.

The device which is to be tested is not a weapon; it is very heavy and it needs much mothering. In its present form it could not be delivered by any ordinary military means. But the fact that it is not a weapon is important only in terms of time; if the device works there will be thermonuclear weapons in a very few years, and compared to this test, the test of the eventual weapon will be a discounted anticlimax. About the so-called hydrogen bomb there has always been this one great question, "Is this possible?" This question will be answered if the project test succeeds.

The memo is undated but carried a cover sheet with a handwritten date of November 5, 1952. The memo can be found in "Foreign Relations of the United States, 1952–1954, National Security Affairs, Volume 2, Part 2" [Disarmament Files, Lot 58 D133]. The memo, which contains a brief introduction not included here, was distributed unsigned by members of the Panel of Consultants on Disarmament. While the panel was chaired by J. Robert Oppenheimer, Vannevar Bush is credited as the driving force behind both the memo and the wider effort to halt or delay the first H-bomb test, according to McGeorge Bundy, who served as secretary to the panel and recounted some of the story in his *Danger and Survival: Choices About the Bomb in the First Fifty Years* (1988), and Barton Bernstein, author of a 1989 paper on the panel entitled, "Crossing the Rubicon: A Missed Opportunity to Stop the H-Bomb?" (*International Security*, Fall 1989).

The test, then, will be a great event if it succeeds. Any such event, in the normal course of administration, is carefully studied by those in authority in order to be sure that it is managed in the best possible way. In the case of this test, however, there is naturally a disposition to believe that the basic decision is past, on the ground that the large questions were those raised and decided when it was originally determined that it was right to try to make a hydrogen bomb. This decision was reached by due process. Should we not regard this test, however striking its results may be, as the natural and routine consequence of the earlier decision? The question is important because Government cannot permit itself the luxury of perpetual self-doubt.

We think that it may be more accurate to conceive of the decision to conduct a thermonuclear test in November as essentially a new decision, deserving the close attention and mature consideration of the highest officers of the Government. We think that much has changed since 1950, and we think also that the very magnitude of the technical accomplishment urges a review of its meaning.

Many relevant changes have occurred since 1950. First, the course of thermonuclear research has modified one set of fears which lent urgency to the quest for a hydrogen bomb. It no longer seems likely that [the German physicist Klaus] Fuchs [who supplied information on American nuclear weapons research to the Soviet Union during World War II] could have been of much help to the Russians in this field since the information he could have supplied them has turned out, in our experience, to be misleading.

Second, we now think we know how to make a thermonuclear device that works, and we also think we can make it into a weapon fairly soon. In 1950, the decision to proceed could not but be stimulated in part by the very uncertainty and ignorance that surrounded the problem; now we know what we are trying to do. The decision to learn about a matter is quite different from a decision to act on what has been learned.

Third, our own stockpile of atomic weapons is very much larger than it was in 1950, and it will be larger still by the time the present thermonuclear device can be turned into a weapon. Moreover, extensions of atomic weapons techniques are making available fission weapons of a yield thirty-fold greater than that of the original bombs; weapons of this size are large enough to deal with nearly all-important Russian targets. While these changes could in large measure be foreseen in 1950, a stockpile on hand is quite different than its impact on thought from one which is merely on order. . . .

Fifth, our experience in Korea and in building NATO has deepened our national understanding of the complex task of resisting Soviet aggression and working for freedom. It is now much more clear than it was two years ago that it is vitally important to distinguish among different kinds of strength and force, using only those which effectively advance our chosen purposes. . . .

Seventh, it has turned out, quite by accident, that if it goes off on schedule, the test will take place either just before or just after Election Day. In either case, it will come in the last months of what the world now knows to be an outgoing Administration. This accident of timing may affect the impact of the test in a number of ways.

Taken together, these changes from the situation of 1950 persuade us that it is proper to raise the question whether or not the projected test should proceed on the present schedule. We turn then to the principal consideration which seem to us to argue for a postponement into 1953.

THE POSSIBLE DISADVANTAGES OF CONDUCTING THIS TEST

1. *The Test Will Assist the Russian Development of a Hydrogen Weapon*

It seems to us almost inevitable that a successful thermonuclear test will provide a heavy additional stimulus to efforts in this field. It may well be true that the Soviet level of effort in this area is already high, but if the Russians learn that a thermonuclear device is in fact possible, and that we know how to make it, their work is likely to be considerably intensified. It is also likely that Soviet scientists will be able to derive from the test useful evidence as to the dimensions of the device.

It may be argued that if we are worried by the incentives which our new discoveries provide to the Soviet Union, we shall have to abandon all research and development. The complaint has force but it is important to observe that the American thermonuclear device is a very special case. First, it is a quite remarkable and complex technical achievement—something of a different order from the ordinary new device which the enemy will inevitably discover for himself in good time. Second, it is a device such that the very act of testing it is public and revealing. Third, national prestige is identified to a unique degree with prowess in atomic weapons. Fourth, and perhaps most important of all, thermonuclear weapons may be far more valuable to the Soviet Union than to the U.S.; this last point is so important that we argue it separately below.

2. *A Thermonuclear Arms Race May Not Be In The American Interest*

Any successful test in a new technical field inevitably accelerates developments in this field throughout the world. Yet such is the character of the hydrogen bomb that we cannot help feeling that the U.S. might be better off if no such weapons existed, even from the immediate military standpoint. The West seems to offer more targets appropriate for such a weapon than does the Soviet Union, and the hydrogen bomb is a relatively more valuable part of one's arsenal if the number of fission weapons available is small; it amplifies the yield of a given amount of fissionable material. Since for the predictable future the U.S. should have a very much larger number of ordinary atomic bombs than the USSR, we conclude that the advantage of a hydrogen bomb is considerably greater for the Russians than for the Americans.

3. *The Test Will be a Barrier Against Work for the Limitation of Armaments*

The U.S. is publicly committed to the notion that the objective of arms reduction is real and important; in the last year, the American government has taken the lead in reopening discussion of disarmament in the United Nations. This policy and these efforts are likely to be prejudiced if the projected tests should be successful, especially as it would explode in the middle of the annual meeting of the General Assembly.

Above and beyond the question of embarrassment to our policy and our negotiators, moreover, there is the fact that the forthcoming test has a special significance in the international arms race. A successful test will mark our entrance into a new order of destructive power, and this is the last point of departure now in sight. There is no other foreseen stage in technical development at which it will be so natural to say "stop, look, and listen." If the test is conducted, and if it succeeds, we will lose what may be a unique occasion to postpone or avert a world in which both sides pile up constantly larger stockpiles of constantly more powerful weapons.

4. *The Test Will Have An Unsettling Effect on Free Nations*

While some in free countries (perhaps particularly in Great Britain) may welcome the November test as an indication of growing American deterrent strength, it seems likely that an explosion of this character will, on balance, be disturbing to most of the non-American, non-Soviet world. It will lend color to the arguments of those who falsely maintain that the U.S. is irrevocably committed to a strategy of destroying its enemies by indiscriminate means and at whatever costs. It cannot but add to the fears of those Europeans who recognize that a poker game played with hydrogen bombs is one in which only the two Great Powers could buy any chips.

5. *It May Have a Hardening Effect on the U.S.*

We think there is a danger lest a preoccupation with destructive weapons should tend to obscure the subtle and varied character of the ways in which we must try to cope with the Soviet Union while avoiding a third world war. We think this preoccupation might be considerably stimulated by the feeling that "we are successfully entering the field of hydrogen bombs." (This sort of loose interpretation of the projected test would seem to us almost inevitable.) To put it another way, we think it important that the balance of action of the American government should have the public meaning that our policy is flexibly designed to cope with both the Soviet Union and the dangers of all-out war; such a balance of action is already hard to achieve, and this test might be a further heavy weight on one side.

6. *The Test Comes at a Bad Time*

Those charged with the responsibility of thermonuclear development have been under urgent orders to develop a hydrogen bomb as rapidly as possible; every priority has been given to this program, and it has been assumed that no consideration of politics or policy should weigh against the need for speed. As a result, the first full-scale thermonuclear test has been scheduled without regard to any considerations except those of making headway toward a weapon. And by accident it happens that unless a postponement is ordered this test will place in November, during the last months of an outgoing Administration. We must unhappily state our feeling that this may be the wrong time for an act of such importance. . . . We are forced to the conclusion that if the consequences of conducting this test are as large as we think, the decision belongs to the incoming and not the outgoing Administration. . . .

Even if it be agreed that the basic problems posed by this great new technical advance are problems which belong to the next Administration, it may still be thought that the basic task of those now holding responsibility is to duck no hard choices, and to continue to act in full responsibility until the new Administration is installed. This position in our view misses the main point. Except in cases of urgent crisis, the great responsibility of an outgoing Administration is to help get the new men off to a good start. The American tradition both expects and honors acts of restraint by executives alive to the claims of those who are about to take on the enormous responsibilities of high office.

THE POSSIBLE ADVANTAGES OF A POSTPONEMENT [OF THE TEST]

1. *A Responsible Government*

If it should be possible to postpone the test with some understanding in the Government and with no great public outcry, the whole matter can then be examined and judged by an Administration fully responsible for the next few years of American policy and answerable for the meaning of its actions. This is the largest and most certain gain which we see in a postponement. A new Administration's decision to test or not to test could have the character of a fully considered commitment to the future in a sense not possible in November.

2. *The Possibility of an Agreement to Abandon Atomic Tests*

Until we have tested thermonuclear devices there remains one opportunity for an international agreement on armaments which would avoid the overwhelmingly difficult problem of disclosure and verification. An international agreement to conduct no more atomic tests could be monitored by each major government on its own. It is a technical fact that no important atomic explosion can take place in the Soviet Union without our knowledge, and there is no reason why the Soviet Union should not develop the same capacity for detecting our tests, if it has not already done so. It is possible to bury a test so far underground that the only thing known about it is that it took place, but this piece of knowledge is all that is needed to monitor such an agreement. Thus an agreement of this character has the unique characteristic that it separates the problem of limitation of armament from the problem of "inspection." Moreover such an agreement would have real meaning, since for some time to come no nation will have any proven thermonuclear weapons if it is unable to conduct and learn from the thermonuclear tests. . . .

We recognize that the Soviet Union might well reject any proposal for the abandonment of atomic tests even if such a proposal were made at a time when it involved a limitation upon American development. The Russians would certainly be suspicious. They might simply denounce the proposal as unfair, since we have had many more tests than they. They also might think the proposal stemmed from American inability to reach a workable design for a thermonuclear device (which might have the effect of moderating their own thermonuclear efforts). They might also respond by trying to entangle the American proposal in their own propaganda for a general prohibition of atomic weapons. In general it is quite possible that the Soviet Union would react in an unconstructive way.

But the important point is not that a proposal of this character might be rejected. The main consideration, to us, is that this may be a real chance to inquire into Soviet intentions and attitudes. So great is the damaging effect of our ignorance of the pattern and content of Soviet Power that we should always be glad to find a topic on which discussion itself may be illuminating. A proposal of this character, seriously and carefully advanced to the Soviet government, should produce valuable evidence of the degree to which the rulers of the Soviet Union understand the character of the race in weapons of mass destruction. It would certainly provide a medium in which the basic American concern with the implications of the arms race can be forcefully presented.

The basic attractiveness of the notion of a standstill in atomic tests is that it offers to the American government something which is exceedingly difficult to find—an opportunity to reinforce its verbal adherence to the idea of disarmament with a visible and measureable action. Such an action might strongly reinforce all those abroad who believe in the good faith and peaceful purpose of the U.S., and it could turn the mind of the American nation itself to the fact that policy in the 1950s must combine strength with moderation and firmness with flexibility. . . .

Plainly, we cannot assert that no risk whatever is involved in any decision which would delay the time at which we have thermonuclear weapons. There is always some risk in any decision to delay the development of any weapon. But in our view the risk involved in postponing the presently scheduled test is not large enough to weigh heavily against the arguments in favor of postponement. . . .

WHAT COULD FOLLOW POSTPONEMENT?

The largest and most difficult of the objections to a postponement is the simple question of what we would do if we got it: what line of policy should we pursue to make use of the time gained? We have argued the claims of a new Administration, but this claim may not be decisive if in the end the new Administration has no other alternative than to reschedule the test as quickly as possible. We cannot avoid the question of what we want to do while this remarkable new device is not being tested. Even if we answer this question by saying that the United States should press for an agreement to prohibit all atomic tests, we are faced by the fact that such agreement will have a fairly short life if nothing else is added. In other words, in order to feel confident about any single step to make disarmament less unlikely, it is necessary to have in mind some reasonable sense of the way in which the whole subject of arms limitation fits into the whole of policy.

Our basic assignment has been to consider this larger question. We have been forced to recognize the strength of the following three propositions which are exceedingly hard to reconcile with one another. First, no limitation of armament is feasible unless it becomes part of a larger understanding of some sort. Second, most sorts of understanding with the Kremlin are either impossible or undesirable or both; we do not know that peaceful co-existence is possible, but it is plain that even if it be possible, it cannot be comfortable or cordial. Third, unless armaments are in some way limited, the future of our whole society will come increasingly into peril of the gravest kind. . . .

CONCLUSION: LET US POSTPONE THE TEST, IF SUCH A DECISION CAN BE UNDERSTOOD, EXPLAINED AND PROPERLY SUPPORTED

Taken together, the arguments for a postponement of the projected thermonuclear test seem to us persuasive. We think that November is not a good time, and we think that the decision should be made by the next Administration. We think that this is a fateful step, and that before it is taken the next Administration should be quite sure that that there is no better use to be made of all we have learned since 1950. We are not persuaded by the claim that postponement would bring unacceptable dangers, and while we admit that it is not clear where a postponement would lead, we have to note that this ignorance applies to any effort to limit the current power struggle. We think the test should be postponed, and though our first concern is with the limitation of armaments, we think that postponement remains desirable when judged from the broad standpoint of the national security. . . .

35

"THE SEARCH FOR UNDERSTANDING" (1953)

In this excerpt from an address at MIT, Bush considers humanity's "search for understanding" to provide its own rewards and justifications, even when conducted against the backdrop of possible nuclear annihilation or other existential threats. Here Bush reveals his own yearning for deeper meaning in a world of science dominated by materialist explanations. He insists the world of rational inquiry is underpinned by ideas and ideals, aspirations and values. "To pursue science," he observes, "is not to disparage the things of the spirit," because the pursuit is an existential calling, a kind of condemnation. "No man delves into the unknown who is not under sentence of death," Bush adds in a striking echo of the wider sense of the human condition that Albert Camus, the French writer and philosopher, referenced in his essay, "The Myth of Sisyphus," which first appeared in English translation two years later, in 1955.

"THE SEARCH FOR UNDERSTANDING"

This is a difficult world in which we live today, and scientific men do well to ponder their part in influencing the way in which we now proceed. But

Bush delivered this address at MIT's Opening School Convocation on October 5, 1953 (MIT Archives, 78/19/Convocation). A different version of this essay, carrying the same title, appeared

I believe that the extreme materialistic view is held only by those who become intoxicated by a bit of a grasp of material things. To pursue science is not to disparage the things of the spirit. In fact, to pursue science rightly is to furnish a framework on which the spirit may rise.

Fear is not new in the world, nor is the problem of evil. The rabbit that crouches as the owl swoops knows terror, and the mother partridge dragging a wing to lure the invader from her chicks does so with a wildly beating heart. And that which is evil to the duck on her nest is good for the fox pups saved from starving. There are life and beauty in the world but fear and beauty are with them, for that is how the world was formed. And man, who gives these things names, wonders at his lot and is baffled and confused.

There was no compassion in the world until man brought it. Nor was there beauty or virtue until he thought it so. His values do not all derive from the will to live or from the sifting of selection. That a man will devote his life to the good of his fellows is not always a product of evolution or self-seeking that is sublimated. Altruism is a product of his mind, not of his seamy history.

So too is his will to know. His yearning to understand reaches far beyond the control of nature for his bodily well-being. The shepherd on the hill at night views the stars and ponders, not just that he can thus care better for his sheep, not just that he is idle and his mind roams, but because he wonders whether, beyond the stars, lies the reason why he can thus ponder.

The search for knowledge has always been under stress. The old geometer, manipulating his triangles in a quest for release from his perplexity by logic, was never far from the barbarian and his spear. The shadow of the guillotine fell across the pages on which France recorded some of its most profound science. No man delves into the unknown who is not under sentence of death. The greatest work of man, a brain trained through the years to deal in wisdom, is destroyed in a moment by chance or malice. The inherited knowledge of the years endures, passed on and accumulated, but even that ends if free men fail.

A plane flies overhead on its peaceful mission. But it may someday be a plane which carries destruction such as the world has not yet seen. The

in *Science Is Not Enough*, an essay collection carrying the subtitle "Reflections for the Present and Future," published in 1967 by William Morrow & Co.

edifices of the city may be consumed in a moment, and with them the edifices of the mind. The threat is very real.

It is well that we should band together in resolve that the deadly plane shall not fly. It is even well that we should strive to ensure that, if it does, we shall still endure. The duty and the opportunity to struggle to preserve our lives, and our way of life, are not cancelled just because the form of the threat has changed.

Do we exist just so that we can struggle to continue to exist? Or is there more to life than that? We understand so little, the universe is so great and so intricate, and beyond it lie things that we shall never know. Is there true significance when man can observe a collision of galaxies which occurred when dinosaurs roamed the swamps or create a painting which conveys far more than is ever seen by the eye? Is there no real value in searching except to use? We are here a company dedicated to the search for knowledge. Are we thus also dedicated to a search for truth? Has the word any meaning beyond convenience?

Because we must fully fail to understand, should we refuse to watch the galaxies as they seem to rush away toward oblivion? Should we probe toward the core of the earth only if it will help us to find new ore? Does man speculate about his origin and his destiny merely because his mind developed the capacity to wonder as an incident to its capacity to help him compete, much as a toucan grew a fantastic bill as a chance product of genes that became combined for useful ends?

Does the mystery of our conscious thought lead us toward a greater mystery, beyond our feeble definition in terms of the marks on rules, or the ticking of a clock, which we cannot understand but which we cannot deny? Are we thus a part of something more profound than the knowledge we gain by the movement of needles on dials or the tipping of a balance? Is there meaning in life beyond mere animal existence?

Science has a simple faith, which transcends utility. Nearly all men of science, all men of learning for that matter, and men of simple ways too, have it in some form and in some degree. It is the faith that it is the privilege of man to learn to understand and that this is his mission.

If we abandon that mission under stress, we shall abandon it forever, for stress will not cease. Knowledge for the sake of understanding, not merely to prevail—that is the essence of our being. None can define its limits or set its ultimate boundaries. Our children's children may weigh it more than we, may be able to lift the curtain just a bit, and believe they know why all this is so. We should leave them a heritage, even though we live in perilous days. Thus we would continue to delve and to ponder, even while we strive to keep the bombs from falling, even if we know the bombs will fall and that things we love may perish. For if we fail to struggle, and fail to think beyond our petty lot, we accept a sordid role. The light in our minds tells us that there is more to life than this.

That the threat is now intense is not a reason to abandon our quest for knowledge. It is a reason to hold it more tightly, in spite of the need for action to preserve our own freedom, in spite of the distractions of living in turmoil, that it may not be lost or brushed aside by the demands of the hour. We would not neglect our duty to our country and our fellows to strive mightily to preserve our ways and our lives. There is an added duty, not inconsistent, not less. It is the duty to so live that there may be a reason for living, beyond the mere mechanisms of life. It is the duty to carry on, under stress, the search for understanding.

36

THE PEAK WAVE OF PROGRESS IN DIGITAL MACHINERY (1954)

In this essay, Bush marvels at the uniqueness of human consciousness, which he describes as "the great central fact of existence." Exploring a theme he's examined over the prior twenty years, he wonders how best to enhance human cognition through computers and other aids to thought. In so doing, he anticipates a major area of innovation in the last quarter of the twentieth century and the first two decades of the twenty-first century. Drawing on his own efforts to imagine a "memex," or memory extender, Bush sketches out a form of "augmentation" of cognition that would later come to inspire leading designers of personal computers in the 1960s and 1970s. The core of Bush's vision is that human reasoning can benefit from "tools for thought" in much the same way that industrial machinery revolutionized physical labor. Computers, he reckons, can free the human mind from mental drudgery, particularly from repetitive tasks. In time, he envisions computing machines that will enable humans to devote more effort to original thinking and deeper insights rather than routine analysis. Ultimately, human minds and computers, he suggests, might work together, building a cognitive partnership that results in unprecedented "feats of analysis" that an individual mind, even an outstanding mind, can't achieve "unaided." While sometimes missing trends in technology, in this area he accurately envisioned a rich future of human–computer relations. In a sobering closing insight, he insists that cognitive enhancement, mediated by increasingly powerful computers, will take place against an existential backdrop of human insignificance in a "heartless cosmos" and that humans will find ultimate consolation in their "search for understanding."

THE PEAK WAVE OF PROGRESS IN DIGITAL MACHINERY

It always gives me a lift to visit California, and to participate in the atmosphere of courage and progress which permeates this part of our great country. It is especially pleasant to meet with you gentleman and to consider with you the subject of research. For research, soundly administered and soundly applied, furnishes a foundation for progress in this modern world; and Stanford Research Institute, in its participation with business on the Pacific coast, exemplifies a type of teamwork which can go far indeed. . . .

I am not going to spend much time on some of the trends which are obvious. . . . let us consider a trend that is not so evident. . . .

There have been great strides in the development of machines for supplementing and substituting for human thought. Digital machines can perform hundreds of thousands of arithmetical manipulations a second, and by so doing can solve even abstruse partial differential equations. There is no reason why man should not relegate to the machine all those parts of his processes of cerebration which are repetitive in nature, or subject to exact formulation, thus freeing his mind for those processes which the human brain alone can encompass. The machine can manipulate, it can remember, it can even learn by experience. . . . But it cannot think in a complete sense. When the machine is joined to man's brain, which can guide it, the combination can perform feats of analysis that reach much farther into the unknown than can the unaided individual, far beyond what can be resolved by groups of men with conventional aids. The digital

Excerpted from the published text of Bush's address to the Stanford Research Institute (SRI) on January 20, 1954. SRI was at that time a unit of Stanford University, and Bush delivered his address at the Statler Hotel in Los Angeles. A copy of the address can be found in Bush's papers held at the Carnegie Institution of Washington.

machine is on the peak wave of progress in analytical machinery, and the analog machines for the purpose are now to some extent overshadowed.

But these are not the only possibilities. In between lie special-purpose devices, such as the crystal analysis machine I have just noted. Someday such machines may enable us to cope with the growing complexities of biochemistry, and bring order and understanding to a field of fundamental importance to the [human] race.

But this process of supplementing man's mind by the machine needs to go much farther than this if we are to cope with the complexities of the world that is coming. Civilization proceeds because man can store, transmit and consult the record, because the accomplishments of one generation are available to the next, because every man can share the experience of his fellows. A great surge forward came when Gutenberg introduced moveable type. Modern methods of printing will soon move beyond the use of metal type and become much more facile. Electrical means of communication, culminating in television, enhance the interchange of current thinking. Modern transportation by air brings men together over vast distances. But we are still in the horse-and-buggy stage in the ways in which we store and consult the record. It looks as though we were about to become buried in a mountain of printed scientific results, with no effective means of penetrating the mass to the find the streak of ore that may be needed for further progress. In the field of physics alone there are thousands of current scientific journals. One of these alone, *The Physical Review*, publishes 5,000 pages a year. No man can handle such masses of material by methods of a generation ago. Scientists, perforce, become limited to the following of highly specialized trails only. Papers with results of genuine value are printed, stored and buried. The conventional library is becoming an anachronism.

Modern mechanization and instrumentation can undoubtedly take us out of this morass. Electronic selection at very high speed, microphotography carried to extremes now possible, facsimile transmission made rapid by television, these and a dozen other techniques can solve the problem. Personally I believe the solution will involve storage and search by association of ideas, in much the way in which the human memory operates. But this great problem of modern methods of manipulating the scientific record is everyone's business, and no one's in particular. Its solution would involve very large sums of money. This is particularly true since pilot-plant experimentation is very difficult and expensive, by reason of the form of the existing record. It is perhaps too subtle a matter to be pursued by government and too pioneering in nature. So it doesn't get done as yet. But it must be solved or we shall be in the fix of a colony of bacteria, bogged down, with all progress stopped by the burden of our own product.

Let me close by emphasizing . . . the search for knowledge which is inherent in the nature of man himself. The motivation which urges scientific men forward is the faith that it is good for man to understand—that this is his mission. The human race is, from one point of view, an incidental excrescence in an enormous and heartless cosmos. We crawl upon a rock which hurtles through space, doomed to struggle with one another in accordance with an inexorable law of evolution, bound for extinction as the earth grows cold, unsung and unremembered, our greatest works but the meaningless scratching on the walls of the prison in which we are confined.

But there is another side to the coin. Man, in his dignity, looks out upon the cosmos and fathoms it. He examines distant galaxies by light which left them a half billion years ago. He traces several billion years into the past, and gropes toward an understanding of the universe as a time-space complex, with a strange geometry which renders it invariant to the point of observation. He constructs his cosmic geometry as the projection of a four-dimensional space configuration, the shadow of which alone he can grasp. He turns inward, breaks down his molecules to atoms, then to protons, neutrons and electrons; watches and manipulates the alterations of these new and strange entities of brief existence. He probes for the ultimate forces which control the form of the nucleus, and thereby control all matter. Moreover, he turns upon himself and examines the mechanism by which nerves act, the mechanisms by which the brain itself remembers, combines and controls. And as he does all this he marvels nonetheless at the great central fact of existence, that he has a consciousness which enables him to comprehend.

It is good for man thus to search for understanding. His search will lead him into dangers, and he may not avoid them. But even disaster, a temporary reversal of the trend which we call progress, is preferable to the monotonous existence of a vegetable. We are embarked upon a great adventure, not of our own choosing or making. It holds perils, and the things we learn may dismay us. But there is a deep meaning to consciousness, even if we cannot grasp it. There is a reason for man's search for understanding which extends beyond his immediate needs for food and shelter. This reason is his inherent urge to know. We share the great adventure with one another. Let us together pursue science for the practical benefits it may yield. But let us pursue it also because we would rise above the sordid daily struggle with nature and with ourselves, look out upon the universe in which we find ourselves, and [go] down into its intricacies. Let us pursue science because we respond to the urge which lies deep in all of us to press on toward greater understanding.

37

"AN OPPORTUNITY WAS MISSED" TO HALT
NUCLEAR ARMS RACE (1954)

In this letter to his closest wartime colleague, Bush expresses grave concerns over the trend toward more destructive nuclear weapons and an arms race with the Soviet Union. Viewing the race as one neither side can win, Bush shares his sense of regret with James Conant, who shared responsibility for administrative and policy aspects of the Manhattan Project. In 1945, Bush and Conant witnessed the first atomic test at Alamogordo, New Mexico; earlier, in fall of 1944, they had co-signed a prescient, if ominous, forecast regarding the American lead in nuclear weapons (see chapter 18, "Salient Points Concerning Future of Atomic Bombs"). Bush wrote the letter on the eve of the secret hearing on the reliability of J. Robert Oppenheimer, whom Conant also knew well; his letter finds Conant, a former president of Harvard and distinguished chemist, serving as President Eisenhower's high commissioner to West Germany.

This letter contains the most complete account of Bush's feelings of regret over what he viewed as the squandered opportunities to halt the spread of nuclear weapons. Even as tests of more powerful H-bombs were planned, Bush held out hope for an agreement to slow, or halt, the race, if only because, in his view, without restraint "we face a very appalling future indeed."

"AN OPPORTUNITY WAS MISSED" TO HALT NUCLEAR ARMS RACE (1954)

Dear Jim:

The H-Bomb explosion has stirred up discussion here as well of course, but not as much as I expected as yet. The situation is that the general public is now beginning to realize for the first time the sort of situation we are moving into, which of course some of us have foreseen for a long time.

I have never seen the report which you mention in your letter of March 26 [the date that newspapers around the world reported President Eisenhower's acknowledgement of an H-bomb test on March 1, near the Pacific island of Bikini] and as you know I was not connected in any way . . . [when] the General Advisory Committee made its studies. At that time there was some discussion in Congress of a separate review on the matter, but I discouraged this on the basis that the General Advisory Committee was set up for the explicit purpose of examining into such subjects, and of course the considerations which reach beyond the technical were presumably being taken into account by the President through his regular advisors. However I do remember well the atmosphere that prevailed at the time and I think there is no doubt that the point of view of the General Advisory Committee has become distorted as far as it has become treated at all.

Ever since the first bomb went off at Alamogordo there have been many of us who have envisioned quite clearly what it would lead to. Now after nearly 10 years we face a situation which will soon be developed to its full threat. We and Russia will face each other over piles of bombs, and it is highly probable that both will have the means for delivering such bombs on targets. It also appears that defense means either group can mitigate the threat by increasing the attenuation of attacking forces, but cannot produce 100 percent protection. The situation is an appalling one, especially on account of the possibilities of surprise attack, but it is also appalling because of what might occur if Russia utilized its potentiality in a series of bluffs. This situation does not come to those of us who have been connected to the matter in one way or another as a sudden surprise; it has been very gradually developed. But I think [the nuclear face-off]

Vannevar Bush to Conant, March 29, 1954 (Library of Congress, Vannevar Bush Papers, Box 27).

comes to the American people as a surprise and a shock, for I think they have been prone to think of atomic bombs only in terms of offense without considering what might happen to this country as events proceeded. It is also true, I fear, that many of those who have been most active in developing our offensive capabilities have emphasized the desirability and effectiveness of a strong striking force, without equally emphasizing the other side of the shield. This is natural enough, I suppose, for those who are trying to get a difficult job done, but it has not produced a balanced point of view in this country, and only now I feel are the thinking people in this country coming to a realization of what is involved.

Now my point of view throughout this whole affair has been a very simple one, and I feel it is the same point of view as is held by most of those who have thought about the subject deeply and from all angles. On the one hand I have felt strongly we must go ahead with full speed with atomic weapon development, for it was fully evident early in the war that this new and powerful weapon was going to arrive in one way or another and that it would be disastrous if we were not equipped and an enemy made a full development. In fact I had a very serious but brief discussion with Mr. Roosevelt on this point one day when it appeared questionable whether our own development could be finished in time to be of use in terminating the war, and his point of view was very strongly that, since this weapon was bound to appear in the world, we must push forward to be sure that it appeared in our hands at least as early as elsewhere. Now, on the other hand, I have felt while this program went ahead vigorously, there should be no lack of effort to prevent the sort of situation that we will now soon face that could possibly be accomplished by international agreement. You remember of course that this was the point of view that actuated all of us at the time that we worked on the proposals which were made to the United Nations. In spite of rebuffs and discouragements, I feel it is essential to maintain this point of view and to seize every opportunity [to slow the nuclear arms race]. This was decidedly the point of view of the panel which was set up by Secretary Acheson and which worked during the summer of [1952] on which Allen Dulles, John Dickey, Joe Johnson, Robert Oppenheimer and I struggled. We did not arrive at any solutions of course, but there was unanimity in the panel throughout and the point of view was exactly what I have expressed, namely that while we advocated increasing the military strength of the United States as rapidly and completely as possible, we nevertheless searched assiduously for a way out of the morass.

I think many have taken the point of view, however, that any international agreements or the like were utterly impossible and that hence any attempt to state alternative to an arms race would be merely futile efforts. I can give, however, one piece of personal experience which I think indicates that the situation was by no means barren of opportunity. I can mention this because it did get reported in the press a bit, in a thoroughly distorted way it is true (although of course no information came from me). It was a personal affair, for at the time that it came up the panel which I mentioned above had not been cleared and put at work. I discussed it with a few people personally and then went to [Secretary of State Dean] Acheson with the matter. My proposal was that in as much as we had not tested an H-bomb, and neither had Russia, since any such test by one could be detected by the other, there was possibility of a very simple agreement, made without fanfare and by direct diplomatic negotiation, to refrain from making such a test at least for a period until further conversations could be held.

Mr. Acheson, I believe, discussed the matter with the President [Truman], but I did not have an opportunity to do so. In fact the attitude in government circles here in this country at that time was such that the proposal simply could not receive calm evaluation. The entire program for the building of an H-bomb was so vigorously underway that any suggestion of delay received practically no consideration whatever. Yet in retrospect I feel that an opportunity was missed and that history will probably record that it was. Of course I have no idea whatever that the Soviets would have entertained such a suggestion, but it seemed to me that the attempt was worthwhile even if it merely showed their point of view. Moreover I felt at the time I made the suggestion, which was some months before the first H-bomb test, that there was time to pursue the idea without indeed causing delay if a complete rebuff [were] encountered. Today I do not see similar opportunity for a specific move, but [then, in 1952] it was quite evident that we still ought to examine assiduously to be sure that no opportunity [was] missed, for the alternative is very grim.

Throughout this affair there has been criticism at times of those who took the dual point of view, and there have even been attacks on those who did so, usually attacks made by individuals who were so sure that their program of building great offensive strength constituted the only sound course that they were utterly intolerant of any discussion which appeared even to suggest that the path of agreement might be simultaneously pursued. But I think this country is now coming to a more realistic conception. The program of vigorous development has certainly been pursued without interruption and I know of no place where anyone has even suggested that it be slowed down in any way unless this be done on the basis of some sort of agreement which by its nature or by reason of inspection or the like could be reasonably relied upon. I hope that the full realization [of the possibility of H-bombs extinguishing life on Earth] does not produce panic or ill-considered moves that might be disastrous. There is some hope in my mind that after we have lived with this sort of thing [imminent annihilation] in the world for a while, the general attitude everywhere may change and agreements of some sort may indeed be feasible. In fact if this does not occur we face a very appalling future indeed. . . .

38

IN THE MATTER OF J. ROBERT
OPPENHEIMER (1954)

*Bush was among the staunchest defenders of J. Robert Oppenheimer, the
celebrated physicist who was the scientific leader of the Manhattan Project.
Oppenheimer supported building atomic weapons and using them against
Japan. After World War II, he served as an advisor on nuclear weapons
to the U.S. government and was chairman of the crucial General Advisory
Committee of the Atomic Energy Commission (AEC). In this role, Oppenheimer disagreed with Ernest O. Lawrence and the influential Edward Teller
on the wisdom of building and testing the "super," which was a hydrogen
bomb that would be significantly more destructive than the original class
of atomic bombs. Other insiders agreed with Oppenheimer, including Bush,
James Conant, the physicists Enrico Fermi and I. I. Rabi, and Lee DuBridge,
president of the California Institute of Technology. Nevertheless, under the
second chair of the AEC, Lewis Strauss, Oppenheimer lost his insider influence over his H-bomb positions. A faction led by Strauss moved to punish
Oppenheimer for his opposition by stripping his security clearance from him,
and thus evicting him from secret policy debates. In support of his move,
Strauss alleged that Oppenheimer was too sympathetic to the Soviet Union
and had lied about past associations with members of the U.S. Communist
Party. Rejecting the option of resigning and advocating for his positions outside of official government circles, Oppenheimer fought Strauss, triggering a
secret hearing about his loyalty in April 1954.*

    *Bush considered Oppenheimer to be an American hero and a first-
rate research leader. Bush privately defended Oppenheimer's loyalty to*

Strauss. In a letter five days after he testified in secret at Oppenheimer's security hearing, Bush wrote Strauss, insisting the proceedings give the appearance that "a man is being tried for his opinions." Bush warned of a situation "where the President becomes accused of thought control." Bush sought to protect Strauss and his allies by keeping his objections private. He viewed the case against Oppenheimer as bogus and believed that the attacks on Oppenheimer highlighted the need for greater protection of dissenters within the scientific community. While he refused to publicly air his objections to Oppenheimer's treatment prior to and during Oppenheimer's security hearing in April 1954, Bush insisted privately that the government's treatment of Oppenheimer was wrong and would instill fear among civilian scientists working on military projects and perhaps reduce cooperation by scientists who might otherwise work with the military. By the summer of 1954, however, Bush broke his public silence and expressed his support for Oppenheimer in a featured interview with Newsweek *magazine. Two weeks later the influential senior editor at the* Washington Post, *Alfred Friendly, reviewing the newly published transcripts of Oppenheimer's security hearing, described Bush as a "leading defender" of Oppenheimer, while in the same breath calling Bush "the grand old man of American science". What follows are several documents, including Bush's appeals to Strauss, excerpts from his testimony to the AEC security board and his Newsweek interview, and a revealing letter to Conant in which Bush hopes "we probably have reached the peak of absurdity." In the end, not only did Bush defend Oppenheimer and assail the methods of his attackers, Bush challenged Oppenheimer's critics to also attack him publicly because, he argued, his own dissent over the government policy on H-bomb testing and deployment mirrored Oppenheimer's own.*

No one ever did.

IN THE MATTER OF J. ROBERT OPPENHEIMER

April 19, 1954: Bush to Lewis Strauss, chair of the Atomic Energy Commission and influential critic of Oppenheimer:

Dear Lewis: Your note surprised me a bit for I could not remember that I had mentioned you in my recent talk with Don Quarles [then appointed

Assistant Secretary of Defense for research and development and later, under President Eisenhower, Secretary of Defense] as we were discussing the problem of relations between the Defense Department and the scientists. And I think you know, from my earlier talk with you, that I am certainly not prejudging anything whatever, and that the fine relations I have had with you for many years have not altered in any way whatever [because of Strauss's attack on Oppenheimer's loyalty]. My whole thought, in these trying days, is to attempt to further effective relations between scientists and their government, which I labored to create during the war years, and which are now in jeopardy.

Let me leave aside the matters I talked with Don, which are decidedly difficult because of public statements that have appeared recently. There is another matter that troubles me exceedingly. It results from the form of the letter from [Kenneth D.] Nichols [an aide to Leslie Groves during World War II and now general manager of the AEC]. It [the problem] did not strike me when I first read [the letter] or I would have commented on it at once and urged a re-wording. Let me state the matter very baldly for that is the way in which [the letter] will be discussed.

One thing we object to in the Russian system is this: When the top oligarchy has decided upon the party line, if a citizen then even expresses doubts his career is destroyed.

Now let us look at the [Nichols] letter. It first states that there is information that raises questions. The meaning is that, if the information reported is true, the individual [Oppenheimer] should be barred from further participation with government on secret matters. Then follows a list of items.

One of these items, which stands alone, is that he strongly opposed a program on various grounds: moral, technical, political.

Now this item does not assert that the opposition was because of improper motivation. It is not definitely linked with anything else. It merely recites that he expressed strong opinions.

The parallel is an appalling one. In this Nichols letter, it is the executive branch of government speaking. It says that if this man had strong opinions and expresses them, this constitutes one reason he should be barred. He is being put through an ordeal which may break him, and his government says that one reason [to break him] is that he had the courage to speak. Whatever else may emerge, we are close to a precipice when this can be an issue.

Of course those of us who would avoid misunderstanding know that this [hostility to dissenting views] is not the point of view of our government. But, unfortunately, this is what the words say in cold type.

In recent years we have had a trend which distresses me, which has done great harm to our national reputation, a trend toward disregarding the basic principles of our democracy in the hunt for subversives. I have looked to the President to lead us out of this morass, to insist on decency with rigor, and now I am confronted with this.

I have thought of writing to the President, for it seems to me that he only can speak the words which will put the attitude of government in proper light. The matter goes far beyond the relations of scientists, although many of them turn to me for guidance and I don't know what to tell them. But I have not sent such a letter [to President Eisenhower] yet. It seems to me that if I point out the problem strongly to the President, I am bound to suggest a way in which he could act to solve it; and this has me baffled. But disintegration of morale is becoming cumulative, and some prompt action to turn the tide seems to me essential. The difficulty, of course, is to do this without interference with orderly procedures.

So I have hesitated. But I am deeply troubled. And not the least of my troubles stems from the fact that I feel that the President, whom I yearn to aid, is being put in a false light which is singularly contrary to his fine nature.

April 23, 1954: excerpts from transcript of Bush's responses to questions before the AEC's Personal Security Board:

What's your view of Oppenheimer's achievement at Los Alamos?
He did a magnificent piece of work. More than any other scientist that I know of, he was responsible for our having the atomic bomb on time.

Was the bomb delivered on time and, if so, what was the significance to the delivery of the A-bomb on time?
That bomb was delivered on time, and that means it saved hundreds of thousands of casualties on the beaches of Japan. It was also delivered on

---

Bush's testimony of April 23, 1954, *In the Matter of J. Robert Oppenheimer: Transcript of Hearing before Personnel Security Board*, pages 560–568.

time so that there was no necessity for any concessions to Russia at the end of the war. It was on time in the sense that after the war we had the principal deterrent that prevented Russia from sweeping over Europe after we demobilized. It is one of the most magnificent performances of history in any development to have that thing on time.

When you became chairman of the Joint Research and Development Board in 1947, did you set up an Atomic Energy Committee?
That is right. I appointed Dr. Oppenheimer as chairman of it as I remember.

In 1952 did you talk with the Secretary of State about postponing the test of the H-Bomb?
I did. That had nothing to do with [the Secretary's panel on the question.] That was a personal move that [I] made, as a matter of fact, before the panel was in operation. The clearances on the panel were delayed. In that interim I visited the Secretary of State and gave my opinion in regard to that test. Before so doing I talked with a number of my friends [including] Elihu Root. I also talked with three or four members that were waiting to go to work on the panel. John Dickey, Joseph Johnson, Allan Dulles. Robert Oppenheimer. I undoubtedly discussed it with one or two others. In every case it was discussing the matter in generalities without going into confidential matters. . . . I then visited the Secretary of State and gave him my opinion on the matter . . . I gave the Secretary of State a memorandum which gave him my personal views. I made no copy of that memorandum. Nobody knows the exact content of that memorandum as far as I know, except the Secretary of State and anyone he may have told about it. It has never been made public. It seems to me it would be quite improper for me to give you the content. I will lean on the judgement of the chairman. My inclination is that I should not reveal this before this board.

Without discussing the memo, can you express your views on the subject as you've done with a number of people?
Quite right. I can readily say what moved me to go at all, and what the general tenor of my thinking was, much as I discussed it then.

There were two primary reasons why I took action at that time and went directly to the Secretary of State. There was a scheduled test [the Mike test, at Eniwetok Atoll in the South Pacific] which was evidently going to occur in early November [1952]. I felt that it was utterly improper—and I still think so—for that test to be put off [conducted] just before election, to confront an incoming President with an accomplished test for which he would carry the full responsibility thereafter. For that test marked our entry into a very disagreeable type of world.

In the second place, I felt strongly that that the test ended the possibility of the only type of agreement that I thought was possible with Russia at that time, namely, an agreement to make no more tests. For that kind of agreement would have been self-policing in the sense that if it was violated, the violation would be immediately known. I still think we made a grave error in conducting that test at that time, and not attempting to make that simple agreement with Russia. I think history will show that was a turning point that when we entered into the grim world that we are entering right now, that those who pushed that thing [the H-bomb] through to a conclusion without making that attempt have a great deal to answer for.

This is what moved me sir. I was very much moved at the time.

Turning now to the fall of 1949 and the controversy over whether or not to proceed with an all-out program to develop the H-Bomb: did you have any official participation in the decision taken at that time?
No I did not. I had no official connection with the matter. I would like to make one thing clear. . . . I think it is fully evident that the hydrogen bomb was of great value to Russia—much greater value to Russia than to us. I think I can also be sure that a test by us of a hydrogen bomb would be of advantage to Russia in the prosecution of their program.

On Oppenheimer's loyalty:
I had at the time of [his] Los Alamos appointment [to direct the secret laboratory to design, construct, and test the first atomic bomb] complete confidence in the loyalty, judgement and integrity of Dr. Oppenheimer. I have certainly no reason to change that opinion in the [intervening years]. I have had plenty of reasons to confirm it, for I worked with him on many occasions on very difficult matters. I know that his motivation was exactly the same as mine, namely, first, to make this country strong,

to resist attack, and second, if possible, to fend off from the world the kind of mess we are now getting into. . . .

I can assure you . . . that the opinions being expressed are my own. They usually are.

On the inquiry into Oppenheimer's trustworthiness, and the insistence of the chair of board of inquiry, Gordon Gray, that "this is not a trial":

If it were a trial, I would not be saying these things to the judge, you can well imagine that. I feel that a serious situation has been created and I think that in all fairness I ought to tell you my frank feeling that this has gotten into a very bad mess. . . .

You can quote me to that effect, I think some of the things we have seen have been scandalous affairs. I think in fact the Republic is in danger today because we have been slipping backward in our maintenance of the Bill of Rights.

. . . . I feel that this board [of inquiry] has made a mistake and that it is a serious one. I feel that . . . this bill of particulars [against Oppenheimer] is quite capable of being interpreted as placing a man on trial because he held opinions, which is quite contrary to the American system, which is a terrible thing. . . . Here is a man who is being pilloried because he had strong opinions, and had the temerity to express them. If this country ever gets to the point where we come that near to the Russian system, we are certainly not in any condition to attempt to lead the free world towards the benefits of democracy. . . .

I think this board or no board should ever sit on a question in this country of whether a man should serve his country or not because he expressed strong opinions. If you want to try that case, you can try me. I have expressed strong opinions many times, and I intend to do so [in the future]. They [Bush's opinions] have been unpopular at times. When a man is pilloried for doing that [expressing his considered views], this country is in a severe state.

———— ∞∞∞ ————

April 28, 1954: Bush to Lewis Strauss, five days after Bush testified at a secret AEC hearing in in defense of Oppenheimer. Frustration mounting, Bush no longer tries to sugar coat his outrage over Strauss' handling of the situation:

Dear Lewis: . . . I do hope you see a way out of the [Oppenheimer] morass. We must not have a situation where the President becomes accused of thought control. And, in addition to the Oppenheimer matter on which I wrote you, there are cases in the Department of Defense that trouble me exceedingly.

When I appeared before the [AEC security board on April 23] I told them, as I felt I had a duty to do, that their entertaining of a set of charges, open to the interpretation that a man is being tried for his opinions, placed them—and hence the [Eisenhower] Administration—in a difficult spot. But of course I have not talked to the press, and I have urged others not to, while the board considers the matter. I also expressed my strong opinion that this country is fortunate that it has citizens willing to undertake such a difficult task as they have before them.

April 30, 1954: Bush to Karl K. Darrow, an American physicist, secretary of the American Physical Society and early author of a book on atomic energy. Despite his insistence on Oppenheimer's loyalty and the damage inflicted on the relations between science and government because of the allegations against him, Bush continues to wish to make his case in private:

Dear Karl: Thank you for the invitation in your letter of April 20 and by phone last evening. I do wish to have you understand quite fully my point of view on this whole matter.

I am anxious to do anything that I can do properly in aiding Robert Oppenheimer in his present terrible ordeal. In fact I have been decidedly active on this matter in the internal relationships within government, and I have appeared before the [security] board. Being thus active, I have felt that it would do a great deal of harm rather than good if I appeared publicly on the subject or expressed opinions publicly, so I have refrained from doing so, and I believe that my effectiveness, if I have any, is best preserved in that way.

Vannevar Bush to Karl K. Darrow, April 30, 1954 (Library of Congress, Vannevar Bush Papers, Box 30).

When Hans Bethe [the German-born physicist, who headed the theoretical division at Los Alamos and worked closely with Teller] first suggested the possible appearance at the Physical Society banquet, I reacted immediately with the thought that perhaps here was something that I could do that would be significant and timely. But on further thought I concluded that here also I ought to be careful not to appear in any way to be bringing pressure on the [AEC's security] board. I hope, of course, that Robert [Oppenheimer] will appear at the dinner just as usual and quite as a matter of fact or regular habit. However I do not think I have ever been to one of these dinners and for me to appear at the moment is not hence in quite the same category. Moreover I have some other appointments and while I would have changed these all about if I thought I could have accomplished anything, I was not inclined to change them under the circumstances.

I hope I was not too abrupt last evening. This whole thing has stirred me so deeply that it is difficult indeed for me to be normal when any aspect of it is mentioned. I always hope that I am taking the right course in trying to help and that you will understand."

June 13, 1954: Breaking his public silence on the government's attack on Oppenheimer, Bush publishes a stinging rebuke to the attackers in an essay, "If We Alienate Our Scientists," in the Sunday magazine of the *New York Times*, accompanied by a sub-headline in which Bush is quoted as saying, "The stifling of opinions can wreck any effort of free men, but it can wreck science more rapidly and completely." The first two paragraphs of his essay, and the final one, capture his conciliatory tone and critical perspective. In the windup, he insists, scientists are "partners in a great endeavor," not "lackeys."

The distinguished board which has reviewed the charges against Dr. Oppenheimer has now reported, and the situation is therefore open to discussion. But this particular case is only part of the whole situation, and it marks a trend which is disheartening in this democracy of ours. It will hence be more useful at this time, and more appropriate, to

Vannevar Bush, excerpted from "If We Alienate Our Scientists," *New York Times*, June 13, 1954.

discuss some of the broad questions which underlie the present dis-
maying situation in which we find ourselves, than to discuss this one
case alone.

This case is not isolated: there are others throughout the government,
not so spectacular and conducted to a considerable extent in secret, and
the atmosphere which surrounds this particular case undoubtedly surrounds also to some extent every action being taken under the present
program of the current security system. It is not possible, of course, to
discuss fully cases which are now secretly under investigation in the Pentagon, where the relationship to our national defense is intimate; but there
are general considerations which need to be kept in mind if the significance of all that is happening about us at the present day is to become fully
apparent. . . .

Scientists need to be used not as lackeys or underlings but as partners in a great endeavor to preserve our freedoms. They need to be wel-
comed, so they will respond in spite of the lure of peaceful pursuits,
because they are inspired and heartened as they join with their fellow
men to preserve their way of life. They need, as do we all, words—and
far more important—acts and deeds to renew hope in troubled and loyal
minds. They need to see, set up on a pedestal, fortified by the strongest
bulwark in executive acts, those principles which we would abandon at
our peril.

⚉⚉⚉

**June 17, 1954: Bush writes to Conant, working in Germany for the Eisenhower
administration, on receiving a printed transcript of the Oppenheimer hearing, to be publicly distributed. In this excerpt of his letter, Bush also shares
his observations on Senator Joe McCarthy and the Red Scare:**

I have no idea what will happen next. It seems to me the [AEC's security
board] missed an opportunity very thoroughly. They could have, had they
been wise, so written their report that they would have made it clear that
they did not try a man for his opinions. Actually their report, the transcript of the testimony and various other things, show that they did just

---

Vannevar Bush to Conant, June 17, 1954 (Library of Congress, Vannevar Bush Papers, Box 30).

that. I have on my desk this morning the full transcript of the testimony. The [AEC's] general manager's office asked permission of those who testified before they released this and of course I gave my permission. But nevertheless it is a strange thing to do for one does not ordinarily assume when he testifies in closed session that this testimony is going to be made public. However, the volume is a massive one, not indexed, and almost impossible to use. I haven't even found my own testimony in it as yet to see if it got reported at all accurately. If it did, it will show that I stated rather strongly to the board itself the fact that trying a man for his opinions is an un-American procedure. I also wrote a couple of strong letters to Lewis Strauss so I felt I was in the clear when I burst into print last Sunday [in the *New York Times* magazine], for I had certainly done my best to take things up internally before writing for the public.

The McCarthy hearings probably end today. The general feeling among the public is one of disgust with the whole affair. The hearings have done no one any good. They have done quite a lot of harm and as near as I can make out no one involved acted very intelligently. . . . In my opinion we will see the end of the McCarthy affair only next November, when I believe we will have a Democratic Congress by a rather wide margin. Certainly the Republicans have done everything they could to lose Congress, for a thoroughly divided party engaged in bickering is pretty sure to be moved out. In all probability, Ike [President Eisenhower] will get along with a Democratic Congress about as well as he has with a Republican one, but after a couple of years of that I doubt if he will run again. The difficulty with Ike as I see it is that he regards himself as a sort of ceremonial President. He apparently does not read much, and the people about him tell him very little. So I doubt whether he really grasps what is going on.

On balance I think that we probably have reached the peak of absurdity and that the tide will now begin to turn. I do not mean that I think the country will become sane over night, but we may see less of excess in the months to come.

I was up around Harvard a day or two ago. What a nice peaceful place it is. . . . I keep seeing people who want to know all about you and I tell them all that you are standing up well and in fact relishing a thoroughly tough job as you always have.

July 12, 1954: Bush, continuing to speak publicly on "the Oppenheimer case," tells *Newsweek*:

I do not wish to comment on the decision in the Oppenheimer case except to say that it does not affect my complete confidence in Dr. Oppenheimer's loyalty and deep devotion to the security of the U.S. Nor can any such decision expunge from the record Dr. Oppenheimer's contributions during the war and since to making this country strong.

We must earnestly try to prevent the leakage of real secrets. We must find any traitors in our midst and throw them out. We ought to be able to do that in ways which will not at the same time do serious injury to ourselves.

The Oppenheimer case is one manifestation of a trend or pattern which is extremely disturbing. There is grave danger that in attempting to conceal secrets we may find ourselves with nothing—or not enough—to conceal. We are in a scientific and technological race. Unless we keep moving on apace, the Russians may get ahead of us. We cannot keep ahead without the full use of our scientific talent.

A whole-hearted, intimate partnership among scientists, military men and civilian policymakers is essential to our future safety. Neither the scientists nor the military men alone can grasp the practical meaning to warfare of the rapid evolution now occurring. During the second world war, an effective partnership was gradually developed. Recently this partnership has been seriously damaged and is being gradually destroyed. Both in the Department of Defense and in the Atomic Energy Commission, there are practices and have been occurrences which discourage scientists and prevent them from making their maximum contribution to our safety. Our internal security system has run wild. It is imperative to our real security that the trend be reversed.

The scientists want no special treatment or privilege in the community, but they do want to work in partnership and under conditions in which their efforts will be reasonably appreciated.

July 25, 1954: Albert Friendly, in an article on the public release of the full transcript of Oppenheimer's security hearing in April, calls Bush the "Grand Old Man of American science" and a "leading defender" of Oppenheimer. Friendly, a *Washington Post* writer who would become managing editor of the newspaper the following year, observes:

Oppenheimer continued to insist that the H-bomb was not the whole answer. The hearing shows him fiercely energetic in projects pushing for continental defense, radar warnings and tactical, as well as strategic, use of A-bombs. It is here that the transcript makes evident the furious and increasing enmity against Oppenheimer by the Air Force, or in particular, the Strategic Air Command. What Oppenheimer was saying was that war plans based virtually exclusively on mass retaliation, by H-bombing Russian cities and installations, were not the be-all and end-all. He wanted, in short, a more versatile, flexible, atomic arsenal. There are hints he was thinking even of atomic weapons to be used against incoming flights of enemy bombers, and perhaps even atomic antisubmarine defense. But mostly he was insisting on vastly more attention to the tactical use of A-bombs and to continental air defense. The transcript shows that those who opposed him considered that point of view as first cousin to treason.

Washington Post, July 25, 1954.

39

SOME THINGS WE DON'T KNOW ABOUT SOLAR POWER (1954)

Bush's directness is on full display when he bluntly declares, "Scientists are grossly ignorant about some of the most interesting aspects of science." In this selection, the ignorance in question involves solar energy, which fascinated Bush as early as the 1930s. "We would hardly go to the trouble of boring holes in the ground to get oil if we could depend on the sun," he declares. How ironic that he delivered his praise for solar energy to an audience at Rice University in the oil capital of Houston, Texas.

Surprisingly, Bush's fascination with the elusive dream of solar energy roughly equaled his interest in nuclear power. His engagement with solar technology represents an apparent paradox: how can the organizer of the Manhattan Project also be one of the earliest American advocates of solar power? Bush devoted this excerpt from a speech to "some things" scientists wish to achieve but may never do so. Here he shows his penchant for poking fun at scientists and politely puncturing the hubris and sense of omnipotence displayed by many individual scientists. His own skepticism toward the inevitability of scientific advance fits well with the tendency of contemporary sociologists and historians to view scientific knowledge as provisional and incomplete.

SOME THINGS WE DON'T KNOW ABOUT
SOLAR POWER

Scientists are a queer lot. This statement undoubtedly will cause no surprise. But I thought it might be interesting if I explored with you one of their characteristics, one that is not generally understood, one that they fully admit in private. Scientists are grossly ignorant about some of the most interesting aspects of science.

When scientists speak to laymen, they often go into rhapsodies about the great things the members of their guild have accomplished. Sometimes they are likely to imply that, if not omniscient, they are rapidly becoming so in regard to the matters within their scope. Unfortunately some scientists have gone further and have implied *general* omniscience on their part, have spoken *ex cathedra* on subjects they know very little about, such as politics and foreign policy. But these are exceptions. Most scientists are modest chaps at heart. The more modest ones do not make speeches. So I would not have you misjudge them. But I still say that on many things in science, they are sadly baffled and confused . . .

We do not even know how to use solar energy effectively. The sun pours on every acre considerably more than enough energy to run a several thousand horsepower-motor continuously, if we could fully recover it. It would not take a very large part of the state of Texas, at this rate, to give us all the power we could use. We would hardly go to the trouble of boring holes in the ground to get oil if we could depend on the sun; and all we need to do is learn to capture the energy being freely poured down upon us, sometimes too freely for comfort. If we recovered only 10 percent of this free energy, we should still do pretty well if the installation did not cost too much. Even as things now stand, I believe that solar energy could be made economical for some purposes. Stanford Research Institute [in California], a group in Wisconsin and others are working on it. A group at M.I.T. recently made an interesting advance on a photochemical method. The Bell Laboratories have shown a photoelectric device which attains the remarkable figure of 6 percent efficiency. In one way or another

Bush delivered "Some Things We Don't Know," at the first meeting of the Rice Institute Associates at Rice University, May 6, 1954. I'm grateful for the assistance of Neal Lane in locating a printed copy of this essay in the Rice Institute library.

I feel sure the job can be done. If we had a truly effective means of capturing solar energy, say by reversible photochemical reactions, we shouldn't need to worry about atomic energy for our future needs.

Nature does the job of photosynthesis. The chlorophyll in a plant, with the help of various pigments, uses the sun's energy to build a host of chemical compounds. We, as animals, eat the product of this process and manufacture from it the chemical to feed our muscles. But scientists can't even take chlorophyll out of a plant into a test tube and still make it work. [Chlorophyll] quits as soon as it is taken from its natural environment. Ewe can neither make chlorophyll nor duplicate its original environment.

A firefly manufactures a couple of chemicals which when brought together give a flash of light. We can make chemicals that do this but very weakly. If we could do it well, a quart of chemicals might run a powerful highway lamp for a month. . . .

These are details, puzzles which await solution, some of which will undoubtedly be solved as a result of the attack now being made upon them. . . .

40

THE FUTURE OF DIGITAL INFORMATION

Storage, Retrieval, Search, and the Construction of

Knowledge (1955)

An early theorist of information, Bush was deeply concerned about how libraries and large organizations would organize and retrieve relevant information in a timely manner that assisted the organization's need to apply knowledge to new and existing problems. The vast expansion of human knowledge, brought about partly by the mass mobilization of American society during World War II, pushed to the forefront an old problem around the classification of disparate knowledge. Bush was among the first to identify the emerging challenge of information "glut," and of the threat of information overload, especially for scientists, intellectuals, and such professionals as lawyers and administrators. Bush viewed the automation of managing information as a primary technological frontier. His insights and perspectives anticipated the computer revolution, the information economy, and the rise and spread of the Internet. Seminal digital innovators, such as the founders of Google and the inventor of hypertext, credited Bush for his foresight in envisioning how information storage, retrieval, and dissemination would radically change and, in doing so, dramatically increase in value, economically, socially, and culturally. Bush most famously sketched out the germs of his information ideas in his "As We May Think" article (chapter 21) in 1945. In this essay, written ten years later, Bush has the advantage of seeing the commercial beginnings of digital computing and the emergence of new technologies of storage. He thought deeply about the technical and conceptual problems of retrieval:

how to find the precise pieces of information that a thinker wanted from
a pile of seemingly undifferentiated information, which, to make the sit-
uation more challenging, was growing all the time. The goal of ever more
precise retrieval of desired information—and the connections, or links,
between data points—remains crucial to the task of making digital infor-
mation systems more useful to human thinkers and more human-like in
their manner of operation. Here Bush shows why, in the 1950s, he was
among the most astute observers of the early stages of an information rev-
olution. Over the years he remains obsessed with the task of identifying
and presenting "the information that we need." And today, more than a
half century later, the destination for information management that he
envisions continues to animate our contemporary goal for search and
retrieval. As Bush observes, "to code our scientific literature or our legal
documents or any other part of our mounting records and thus to place
them under the control of machinery responsive to our will is a stupendous
undertaking."

THE FUTURE OF DIGITAL INFORMATION

The progress of our civilization in peacetime depends, and has always
depended, not only on our current thoughts and findings, but on the skill
and facility with which we create, store, interchange, consult and utilize
the whole record of our collective past experiences. We are making enormous strides in the development of methods for creating a record of what
we learn—in printed words, by photography or on a magnetic tape. We
are also making strides in developing means for the transmission of ideas
from one to another or from a central point to great audiences. But in one
exceedingly important phase of the whole problem, we are making little
progress indeed. This is the phase of finding in the record the information
that we need. If the record of our experience is to serve us well, we need
to be able to extract from it at will, promptly and inexpensively, any single item of current moment. This is of great importance for the progress

Excerpts from the text of a speech by Bush to the American Society of Mechanical Engineers, on
Feb 16, 1955 (Carnegie Institution of Washington, Vannevar Bush Papers).

of science, but it is also important in every professional field of activity and especially engineering, which is daily becoming more complex in the ways in which it applies science economically to meet the needs and desires of mankind.

We are building the record at a prodigious rate. Books, magazine, technical journals, reports are being produced by the ton. The Library of Congress reported in its *Quarterly Journal of Current Acquisitions* for August 1954 that in fields of science, technology, medicine and agriculture, it received approximately 30,000 journals, including 2,000 new titles; 25,000 research reports; 15,000 books and monographs; 15,000 manuscripts; 10,000 pamphlets; 5,000 prints, blueprints, microfilms; and 150,000 maps and charts.

Our libraries are filled to the overflowing, and their growth is exponential. Yet in this vast and ever-increasing store of information we still hunt for particular items by horse-and-buggy methods. As a result there is much duplication and repetition of research. We are being smothered in our own product. While we record with great care the work of thousands of able and devoted men, full of significance and timeliness to others, a large and increasing fraction of their work is for all essential purposes, lost simply because we do not know how to find a pertinent item of information after it has become embedded in the mass.

The problem is not essentially one of techniques. It is rather one of deciding who is to do the job of clearing up the confusion and under what auspices. There are plenty of good techniques available. Recently I participated in a study of how mechanization could be applied to the problem of searching in the Patent Office, where millions of items need to be scanned for equivalents of the combinations presented in patent applications. This study showed that there are at least 50 companies active in the production of one kind or another of data-handling equipment. I have not time to review all the ingenuous devices now being used for commercial purposes by banks, insurance companies and other businesses, or for scientific computation and analysis in hundreds of fields of research. Suffice it to say that there are several ways in which items can be scanned at the rate of a thousand a second, selected in accordance with a complex code, and reproduced automatically. Photographic methods can reduce the size of a record by a factor of one hundred to one or more, and re-enlarge on call with negligible loss of legibility, thus cramming the material of a thousand books into the space of a cigarette package. Digital computing devices can manipulate records in the form of numbers at the rate of a million operations a second if necessary. Magnetic tape can receive any kind of data, combine them, and reproduce them; and it lends itself to ready erasure, replacement, and rearrangement. There is no lack of powerful, versatile equipment, which is quite capable of rendering our stored records available in prompt, accurate, effective fashion, and at a distance if this is desired.

But to code our scientific literature or our legal documents or any other part of our mounting records and thus to place them under the control of machinery responsive to our will is a stupendous undertaking. Worse than that, it is everyone's business and the assigned responsibility of no one group in particular. There is not the slightest doubt that it would pay, in a very practical sense, to do the job no matter how seemingly great the cost. It would pay everyone and the expense should therefore be borne by everyone. Thus it is a task for government. But is it possible for a government which cannot even mechanize its postal system to be farsighted and courageous enough to undertake a task of such complexity and such magnitude? Probably the methods to be adopted, moreover, are not at all clear at the moment. The whole art of data handling is improving every day, and it would be a mistake to freeze upon a single system prematurely. So it probably would not be wise to plunge in at once and undertake a comprehensive and expensive program for, to use an example, the entire bulk of scientific literature. The various possible approaches need rather to be tried out on a more modest scale. The time has most certainly arrived when special sections of the record can be subjected to mechanization with genuine benefit to those who use them.

Looking forward, I am confident that we shall see further advances, most interesting ones, throughout the whole range of devices by which man communicates with his fellows. And as a result I believe we shall, as a race, if we do not commit suicide by indulging in total war, advance in knowledge and understanding and perhaps also wisdom. I believe we shall advance in our mastery over the records we create, rendering them easier to consult by means which would now seem strange and bizarre to us, which will make obsolete much of what we now do, but will give a new power and freedom to the creative mind and thereby open the way for another spurt forward of civilization. For civilization advances only as it acquires new experience and only as it makes its experience available and useful.

41

FAITH AND SCIENCE (1955)

The historian of science Thomas Kuhn, in his seminal book, The Structure of Scientific Revolutions (1962), *created controversy by comparing the belief systems of scientists with religious belief systems. Explanatory frameworks in natural science, Kuhn argued, could function as a reigning orthodoxy, with a prevailing "paradigm" rejecting or ignoring contrary evidence—with a near religious fervor—until the evidence becomes so compelling that a new paradigm, or belief system, replaces the old one. Though he appears never to have engaged Kuhn directly, Bush insists in this excerpt from his final annual letter to employees and supporters of the Carnegie Institution of Washington that science and research depend on a kind of belief that resembles religious or spiritual faith. "For the scientist lives by faith," he writes, "quite as much as the man of deep religious conviction."*

FAITH AND SCIENCE

Why do we pursue our course of fundamental research at all?

That our motivation is strong we have ample evidence, but we seldom attempt to analyze it. We follow our course so assiduously that we do not

---

Excerpted from *Report of the President* (1955), Carnegie Institution of Washington.

stop to consider where it leads or why we should pursue it. We tend to keep our eyes trained too constantly on the immediate steps before us and fail to look at the fine country that spreads away from us on either side. Though the areas of endeavor that surround us are often inspiringly beautiful, we ignore their long vistas or glimpse them but faintly. They are the goals of others, who with equal dedication and intensity pursue different courses from ours. Our road is that of fundamental science. One does not saunter along it with one's head in the air, for it is strewn with rocks and briars to catch the unwary.

Yet as we follow it and climb its steep ascent with joy and eagerness, there are many side roads by which we could leave it, and these often lead temptingly down into fertile valleys where a less demanding life beckons, furnished with an abundance of all creature comforts. There would be no stigma upon us if we turned off on these spurs, and yet very few of us do. As a company, dedicated to a way of life, we are nearly unanimous in our acceptance of its customs and procedures. And our . . . company includes research assistants who wash a thousand petri dishes or solder a maze of wires, with only a vicarious participation in the grand elation of accomplishment.

What is it that joins us thus in common effort, and urges us on?

A cynic might give very simple explanations. Andrew Carnegie left a fund to be spent in this way; and we, perhaps, find the spending pleasant without giving much thought to our purpose. There have always been aspects of scholarly work that resembles a gentleman's game, made exclusively by a special jargon—Latin or some modern equivalent—and perpetuated by the presentation of a solid front of mutual acclamation in dealing with sources of support, however self-critical it may be within the closed circle of its participants. But no such cynical explanation will suffice to account for what we have here: a group of strong and able men capable of deriving acclaim and reward from the world in other ways, who yet choose to pursue the path of science, though the individual steps of their advance are often appreciated only by their immediate associates and the pecuniary rewards for their effort are small. We must look deeper for true motivation.

Our field is fundamental science. And by this I mean that we seek out the facts of nature and their interrelationships, without attempting to apply them immediately to the needs and desires of man for physical comfort, power to control the forces of nature, or freedom from disease. We seek to acquire and correlate knowledge, not to participate in its application. Long ago we learned that, if we take our eyes of the main trail and view with too much interest the tempting side paths that lead to early practical application, we are likely to lose our way and to waste much time in struggling to find it again. Yet the fact that there are side paths for those who would follow them, and that our blazing of the main trail makes possible a digression into those byways where results helpful to material progress can be secured, is one true and worthy motive behind all we do. We know there may be a long interval between our work and its practical application, that the disconnected relations we find must often be joined with those found by many other investigators before a useful pattern emerges. And yet this does not trouble us; for we are conscious of the problems that the race will struggle with long after our work has been merged in the general advance of science, and later generations will finally complete work we have begun.

Thus we work on fundamental problems of genetics or embryology, the synthesis of proteins by bacteria, or the chemical nature of a virus. We are not concerned directly with the current improvement of domestic animals or plants, the practice of obstetrics, the biochemical production of antibiotics, or the control of virus diseases, even though some of our findings prove to be immediately applicable to all these fields, and we rejoice when this happens. The practice of medicine is still largely empirical; but some day it will rest on a more orderly base, and we would contribute to that achievement. The mechanisms of inheritance are just beginning to be understood in part; and we know, from the mysteries that confront us, that our theories will become far more elaborate and intricate before they become fully comprehensive. One of the greatest mysteries of life is how a minute chemical aggregation, controlling the division and diversification of cells, can produce the intricate, highly adapted organisms which we see all about us—how, in fact, it can produce an organism such as ourselves. We know that if we understood the process by which growth is controlled, much good might follow, including perhaps some control of the process when it becomes degenerative. Our whole lives are affected by the interplay between our organic structure and a host of simpler organisms, some of them beneficial, but most of them predatory and destructive. We are just now learning how these simple organisms operate, and learning that they are not so simple as we thought. The path toward a complete grasp of their biochemistry and structural organization appears to be a long one, which offers many fascinating opportunities for research along the course. And beyond the study of all other mechanisms lies the problem of man's mind itself, its chemical, electrical, and structural aspects. The study of this problem, as well as of theoretical and experimental psychology, is basic to any scientific investigation of the subtler regions beyond, which may include phenomena that are explicable in no familiar terms. The subtler region must be understood more fully or at least defined if we are ever to truly know ourselves, our possibilities and our limitations.

As we regard the present material progress of man and his probable future, we can see that he is headed for catastrophe unless he mends his ways and takes thought for the morrow. This is quite apart from the immediate question whether he will use the split atom or a trained virus to turn civilization back and force it to begin again its slow upward climb—apart from the question whether great wars can be avoided through a general understanding of their consequences. These are momentous questions, but they are immediate ones, and so they must be resolved by other patterns of thought than those of long-range fundamental science, which—though it may profoundly influence the fate of future generations—can affect the present course of events only rarely. They are questions in which all scientists are deeply interested, but primarily as citizens and members of the community—questions to which they can contribute only as citizens or as their background in science enables them to counsel wisely with those who apply science and those who apply other disciplines, all of whom today find some pertinence of science to their problems.

Wars aside, man is still headed for trouble . . . [so we might understand if] the sole motivation of fundamental science [is] to prepare the way for later useful applications and thus fend off . . . catastrophe. And do we always have [momentous goals] clearly in view? Of course not. A great deal of research is done as a matter of habit, because it is the fashion of the time or the practice of the community. We seldom stop to think of why we do what we do; in fact we seldom think philosophically at all. Many a research program is conducted in which the researcher has no clear idea as to what he is trying to find out or what it would mean if he found it. Men work for years without stopping to examine critically the relevance of their efforts. Curiosity is often the underlying motivation, and not a bad one. It was probably responsible for the discovery by Prometheus of how to make a fire. Emulation is often clearly present, especially when a great scientist is surrounded by disciples. The competitive spirit, fortunately now on a more friendly basis than in the past, has a strong part in scientific motivation. It has become less artificial and less acrimonious as caste distinctions and false pride have lost their hold and as the stuffed shirts among us have dwindled almost to extinction. And the sporting instinct, whatever that is which causes men to pit themselves against great odds, to revel in trial and adversity, to breathe the stimulating air of self-justification in success, no doubt is often present. We do what we do for many unavowed reasons, and seldom pause to analyze them. But when we do reflect, we find that our primary motivations in scientific effort extend far beyond our casual and momentary reasons, even beyond the thought that what we do may, in its small way, benefit the human race in its struggle to control its environment and itself in the grim days that are sure to come.

For the scientist lives by faith quite as much as the man of deep religious conviction. He operates on faith because he can operate in no other way. His dependence on the principle of causality is an act of faith in a principle unproved and unprovable. Yet he builds on it all his reasoning in regard to nature. This is even truer today than it was some 60 years ago when William James so forcibly reminded us of it. Some of us, confronted with the bizarre as we delve into the nucleus of the atom or try to pin down the elusive electron, wobble sometimes in our basic faith that natural events are determined by unique causes, that, from a physical standpoint, the present fully determines the future. We become confused as to whether, when we include ourselves as part of the system, causality is departed from. But we still proceed in our daily round on the assumption of unique cause and effect as the basis for all our experimentation. It is becoming apparent also that the acceptance of any system of logic for reasoning about our sensory experience is an act of faith, and that our reasoning appears sound to us only because we believe it is and because we have freed it from inconsistencies in its main structure—for [our reasoning] is built on premises which we accept without proof or the possibility of proof.

42

WHY DO WE PURSUE SCIENCE
AT ALL? (1955)

On the tenth anniversary of Hiroshima and in midst of terrifying expan-
sion of hydrogen bombs by the United States and Soviet Union, Bush
reminds readers that science and the quest for engineering knowledge
extends far beyond the horizons of war and national security. He shows
confidence in the capacity for scientific inquiry to expand humanity's
knowledge of the planet. However, growing knowledge coevolves and coex-
ists with the use of science for destructive purposes. By today's standards,
Bush's insistence that "man now controls his destiny" and that "his power
is growing" seems excessively optimistic. At the same time, Bush resists
"scientific arrogance" and ponders "the limits" on scientific inquiry. Once
more Bush grapples with the challenges of sustaining faith in an age of
science while recognizing the limits of rationality. He insists that "the seat
of ethics is in our hearts, not in our minds."

WHY DO WE PURSUE SCIENCE AT ALL?

Why do we pursue science at all? What are the motivations of scientists? It
is well to examine these from time to time, but especially so as we attempt

Excerpted from the fourth and final section of Bush's essay, "For Man to Know," from the August
1955 issue of *The Atlantic* magazine.

to consider the trends and the ways in which scientists may be called upon to work together in new relationships.

Much of the motivation is clear and immediate. We need not consider applied science, conducted for commercial or military reasons. Nor do we need to consider basic science pursued because it may later contribute to profitable applications. There are more important categories. One is the research aimed at bettering man's lot, without thought of gain: the attack on disease, or ignorance or overpopulation. Here the immediate objective is clear, though the underlying motivation is usually unexpressed. But there is a category of research that has no relation to material advantage; it is fundamental science, directed merely at the extension of man's understanding of himself or his environment, or at the extension of his ability to understand.

The motivation behind much of this kind of scientific inquiry is, no doubt, mere curiosity—that strange characteristic of man which, more than anything else, has led to his ascendancy on our planet and which drives him still toward the mastery of what may now seem the unknowable. Joined to curiosity, and responsible for some of the most amazing flights of genius, is the same aesthetic urge that leads to great art and music. This is peculiarly the case with respect to some of the more abstruse developments in the fields of astronomy, physics, and mathematics. And many scientists derive their strongest motivation from religion and carry on their mission by faith.

The number of youth entering on a career in science today is great, and the interest of the public at large in the work of scientists has grown enormously. What do we say to the young man who is immersed in science, with his whole life wrapped up in it? The new things that he must master, if he will advance with his fellows beyond superficiality to creative activity in a specialized field, are so numerous and difficult that he has little time to spend on the philosophical thinking of the past. He is prone to acquire his philosophical orientation accidentally or casually from his scientific teachers, and in doing so he may be easily misled. For I fear that some of the things that are told him in this respect by scientists are erroneous, logically unsound; yet they are likely to be accepted by him just because they appear in scientific dress.

Science has been enormously successful in its representation of reality. Many subjects, once thought to be beyond its purview, have finally yielded to its attack. [Science] normally recognizes, as it proceeds, that there are things it will never know, things that lie permanently beyond the weak sense of man, however extended by instruments. Yet its scope is the whole panorama of the physical universe and all the physical aspects of organic life on earth; and these vast areas of knowledge, it is confident, will one day be subjected to its dominion. And out of this confidence, too vaguely defined, there sometimes grows a degree of scientific arrogance. Scientists forget momentarily the limits which science initially set for itself.

Let us consider a presentation that is being made today very forcefully and convincingly. There is nothing fundamentally new about it, for the same presentation appears throughout the history of human thought; but it now acquires a completeness and elaboration which are highly attractive to those who think logically, or believe they do. In substance the presentation goes as follows:

We find ourselves in a mechanistic universe, riding on a fragment from a primeval explosion, projected into nothingness, destined to plunge through space for a while, and then to cool to utter inertness. On this fragment evolution has occurred, an entirely mechanistic evolution from the chance appearance of reproducing molecules, through a myriad of species, sorted out by natural selection, to the appearance of man. Now man, the highest animal, is destined to ride for a while and then perish. There may be millions of other fragments, with organic life on them, sentient beings, conscious of their presence in a role not of their own choosing, riding also to their deaths.

We need not quarrel with this presentation so far. Its formulation is one of the tasks of science. In it, science confines itself strictly to the things that can be measured and recorded. It carefully skirts all questions that are not answerable by its methodology.

But from this presentation of the mechanistic universe some recent writers have gone on to formulate a code of ethics as though it followed in logical consequence. This code, in summary, is very simple. Man controls his destiny; let him so control it as to build for himself a better life. That is good which leads in this direction. The code is laudable enough as far as it goes; but it is incomplete and without a logical base in the facts from which it purports to be derived. For it is based on a tacit assumption that the mechanistic account of the universe that has been constructed within the accepted limitations of science is in fact a complete account, and a proper basis on which to build a complete ethical code.

This is to assert that there is no reality beyond those things which we can measure with a rule or time by a clock and that value can be deduced from a statement of fact. But man's motivations emerge from his entire experience. The seat of ethics is in our hearts, not in our minds.

Yet there must be motivations for the moment; we cannot wait until our children perhaps lift the edge of the veil a bit in order to orient our lives today. Thus there is no fault to find with those who go no further than to consider: here we are; let us build a better world—provided this is stated merely as a working formula. Upon such a simple formula, joined with an equally simple definition of what would be a better world, can be erected an acceptable code of ethics. Such a code has been erected by a large portion of those who work in science. It is at times as fine a code as any that has been adhered to by serious men to guide their lives.

Within limits man now controls his destiny. True the processes of evolution are still effective, and man's control is as yet feeble. But his power is growing and can be consciously enhanced. By this power let us build a world in which men may live happily. Let us conquer disease and banish the causes which distort men's minds. Let us end war and the pressure of population upon the means of existence, which is one of its causes. The construction of an environment in which there is peace and harmony is worthy of our best efforts. It may enable a little child to be happy and healthy and to avoid the mental distortions which produce distress. In it there will be music and art and a flowering of those attributes of the race which somehow transcend the dull problems of food and shelter. In it, man may develop his individuality and rise to true dignity. Let us so build that our children may lead better lives. Our code will be a simple one: that is good which leads towards this consummation.

It is dangerous, however, to regard such a code as logically derived from our scientific attainment. The same logic and the same attainment could equally well support a far different conception of what is good. It does support motivations today which are in stark contrast with those by which we wish to live. Logic alone can lead one toward a code which subordinates all means to their end, means of cruelty and deception to an end of regimentation where the free spirit is at the mercy of a communistic state. The urge to serve one's fellow man is not based on any scientific dogma, and the attempt to give it scientific justification involves a danger-ous fallacy. The urge is based on a deep-seated aspiration of the [human] race, which is [our] only hope.

The simple creed of service set forth above suffices for the day's work of many a scientist. Yet there is joined with it a deep conviction, a faith if you will, which for many a man furnishes motivation and satisfaction, entirely apart from the current struggle. This is the conviction that it is good for man to know, that striving for understanding is his mission. We are embarked upon a great adventure, and it is our privilege to further it. Even though at times the box that is opened be Pandora's, even though there are both good and evil in what we learn, it is our duty and our call-ing to extend man's grasp of the universe in which he lives, and of himself. By this process, of beginning to understand, we have made such progress as we have. Though the path be thorny, this is still the way in which we should proceed if we would finally emerge from darkness into the light.

43

THE PIONEER (1957)

In this speech delivered on a visit to the state of Oklahoma, Bush conveys a wise weariness about grand challenges facing Americans. He's trying to convey an enduring message: Nothing of great value comes easily. Speaking less than four months before the Soviet Union launched its Sputnik satellite and ignited near-panic in the United States, Bush highlights the benefits of improving scientific literacy at all levels of American society and increasing the education of scientists. He invokes the words "pioneer" and "frontier" in the hopes of connecting with his audience, some of whom are descendants of land-rushers who came to Oklahoma in 1893 to take advantage of the giveaway. (Bush showed no similar impulse to connect with Native Americans dispossessed by the westward expansion of European immigrants).

Since Oklahoma's two main universities were not research oriented at the time, Bush speculated about the possibility of future research collaborations that would address societal needs that elite researchers might miss. While short of concrete examples, Bush presented an appealing link between research, the geographic frontier, and the figure of the pioneer in American life. Through this juxtaposition, Bush seeks to bridge the gulf in between an educated elite and less-educated masses; in his mind, both segments of Americans can appreciate the excitement and the promise of the new frontiers brought about by advanced research and discovery. So compelling to Americans was frontier imagery that John F. Kennedy, in his 1960 acceptance speech at the Democratic Party's national convention, invoked the term "New Frontier" to refer to his ideas on government renewal. Kennedy declared: "We stand today on the edge of a New Frontier, the frontier of the 1960s, the frontier of unknown opportunities and perils, the frontier of unfilled hopes and unfilled threats." And then echoing statements by Bush years before, Kennedy added: "Beyond that frontier are uncharted areas of science and space. . . ."

THE PIONEER

It would hardly be possible to come to Oklahoma without giving thought to the role of the pioneer. For this was a country of pioneers, not so long ago, and the spirit still remains. In thinking of the pioneer, however, I have some thoughts on his position and contributions in the modern world, and I would like to trace these with you.

I am thinking of course of men and women who are pioneers by choice, not of necessity. It was they who opened up the West, who showed courage and resourcefulness and above all independence as they brought a new land into use and built their homes.

The pioneer differs from the entrepreneur or the promoter. These latter two sometimes contribute substantially to progress, when their work is constructive in nature and not merely parasitic. But they are not pioneers in the sense that I wish to use tonight. For the entrepreneur and the promoter are looking for immediate results, in the organization of new industries, or the acquisition of new minerals, where the object is profit, and incidentally often a benefit to the region in which they operate. But they are seeking a short-time advantage of some sort.

The pioneer on the other hand takes the long view. He struggles and builds, not for himself primarily but for his children. It was this spirit that fired each of the best pioneers who opened up the West. He knew of course that he was not going to an easy life, that his lot would be hard and

Bush gave this address on June 17, 1957, to the Frontiers of Science Foundation of Oklahoma, a civic group in the state dedicated to promoting science education in public schools. A typescript of the text found in Vannevar Bush papers, Carnegie Institution of Washington.

the vicissitudes many. But he held the vision of a sweeter land and a more satisfactory life, and held this vision in terms of what he was contributing to posterity, to his own children and those of his neighbors. This was the finest type of pioneer and his primary characteristic was that he looked beyond the affairs of the moment to a remote and very worthwhile goal.

The old geographical frontiers have now largely disappeared. There are several new frontiers, and most notable among these are the frontiers of science. And here also we need the pioneer, to take the long view, and to seize upon the opportunity that is overlooked or bypassed by those who pursued the well-trod paths. It is the function of the pioneer in science that I would discuss. And before doing so it is necessary to say something about science itself.

We customarily divide science into pure, basic or fundamental science on the one hand, and on the other hand applied science which merges into engineering. In this country we have done well in applied science and in particular in novel and ingenious applications of scientific methods, and in great engineering works which depend upon advanced scientific techniques. But we have not done nearly as well in basic science, and only since the war have we made strong efforts to attain a position in basic science of which we can be proud, for prior to the last war we leaned on Europe overmuch. The great discoveries and the great advances came from Europe to a far greater extent than should have been the case. This was partly the result of our national character, which is enamored of the practical and concrete, rather than the abstract and philosophical. Today we are slowly learning a lesson in this regard and beginning to furnish to basic science the generous support which its importance to our culture and to our future prosperity warrants.

The basic scientist certainly builds for the long future and in that sense he is a pioneer. He looks forward to the day when the manifold contributions of himself and of his fellows will result in the grand synthesis which will bring together for the admiration of all who survey it a new grasp of the order of nature and a new intellectual achievement for mankind. Still his satisfactions and his rewards come from the day-to-day accomplishments, even though his eyes are on a future goal. This is one type of pioneering in science and a very admirable one. But I have in mind another form. And for this other accomplishment an individual does not necessarily need to be a basic scientist, in fact he does not need at times to be a scientist at all.

The building of edifices of scientific knowledge goes on apace all about us. Thousands of keen young men are delving into the intricacies of the unknown and are beginning to understand relationships so bizarre that they cannot be expressed in any language to which we have become accustomed. Undoubtedly at some time in the future this gathering knowledge will lead in turn to new and broader generalizations, then to applications and finally, in some instances, to ways which will improve our health as a people, our living standards and our general well-being. A generation and more of effort will be needed to bring some of these matters to fruition . . . Still, it is mot the only type of pioneering which the frontiers of science afford.

There is no lack in this country of the tendency to seize upon the advances of science and to apply them in useful ways. We can be sure that this aspect of science will not be neglected. With the enormous increase which has occurred in the past decade in research in industrial establishments, with the great growth in the number of able scientists in this country, and of engineers who understand their language and who have the knack of bringing together scientific advance and economic understanding, we are sure to see advances in new instruments, new materials and metals, new machines and new industries. Those who thus bring our scientific progress into use perform a valued function for all of us, even though some of their efforts are ludicrous and some are trivial or worse. In general they have been responsible for a major part of the advance in material wealth which this country has enjoyed. The organizations which apply usually depend upon others for the source of their basic new knowledge. True there are exceptions. Some of the advances in fundamental science have come, we can be proud to say, from our industrial laboratories, where there had been wisdom in combining basic and applied science in a single group, and where scientists have been given their freedom, often as fully as in the great universities. I trust we shall see more of this. As we become more mature in this country, and as we place basic research in its true light, we realize that only by vigorous attention to it can we maintain our position as a strong virile nation dependent upon science and its applications. We may expect this progress in the useful application of science to continue. But in general this is not pioneering in the sense of the word as I use it tonight. For pioneering takes the long view, and these efforts at interpreting and commercializing scientific advance are generally aimed at the immediate and transitory.

Now there are certain areas of science and its applications which are seldom concentrated upon, either by the basic scientist, or by those who apply. These are areas which are intermediate between two general forms of science, and they also areas in which the incentive for either group is relatively lacking. On the one hand, as far as they involve extensions of basic science, they do not have the lure and mystery of more fundamental endeavors. They do not attract young and competent minds as does for example the exploration of the nucleus of the atom. On the other hand they do not attract applied scientists from industry, for there are no immediate applications apparent which promise to pay off the cost and effort which investigating them and bringing them into practice would involve. They are areas in which the long view is highly necessary, and which involve matters which do not now excite vigorous and widespread effort but which some day will be of vital practical importance to the people of this country. It is in these areas, exceedingly important but remote, that I would welcome the efforts of a new type of pioneer. . . .

Although there is fine scientific work being carried out in government organizations, and soundly managed governmental support of long-range research in universities and scientific institutions, I do not believe we can depend upon government to fill the gap, and ensure adequate attention to some of the critical problems of the future applications of science. Some of the programs I have in mind will require many years of intense effort, gradually and patiently accumulating data and understanding, yet constantly calling for vigorous imaginative effort, with a devoted group working doggedly toward a future goal through years of discouragement, with only rare brilliant results to excite admiration and attract support. These programs will also call for drive and action, experiments that are sheer gambles, unconventional methods of approach, such as are more commonly associated with industrial undertakings than with governmental programs. The great governmental laboratories can help, certainly, but I do not feel we can depend upon them to carry the entire burden.

There is a definite place for the independent pioneering group in this field—on the critical problems of the future which are in danger of being relatively neglected. The problems are large and varied and there is need for the impact of diverse backgrounds and still more need for continuity of effort, by groups and individuals, privately supported, devoting their attention to long range problems of great potential importance. And I feel this because I believe that these areas of effort call for the versatility, resourcefulness, courage and above all patience of independently constituted groups. I hope and trust we may expect government to do much in these areas in the future. But government will do still more if there are independent organizations with all the aggressiveness that this should imply, setting the pace and seeking out the opportunities. . . . In fields such as this, which are the concern of everyone, but the immediate practical concern of none, where government proceeds haltingly or not at all, it seems to me that there is an opportunity for a modern type of pioneer, one who can take the long view, one who has a keen urge to do something for those who will follow him on this earth, a pioneer of courage, resourcefulness and ingenuity, and above all a man who can work in a group effectively, and who is nevertheless thoroughly an individual.

In thinking of this visit to Oklahoma I have of necessity thought of the early pioneers. I have thought also of the pioneers of the present day. What could be more natural than in this region where the atmosphere of pioneering persists strongly . . . what [is] more natural than that here in Oklahoma there should be groups leading the way, for the benefit of the country as a whole, in examining into some of the problems of its future which are neither pure science, with all its subtle lure, nor economically attractive for immediate application, but which may hold the key to our prosperity or our failure in the days to come.

44

"THOSE WHO TALK FREQUENTLY BECOME IGNORED" (1957/1959)

In his prime, Bush talked with prominent journalists. Once television news went national, he also appeared live on the air. This short note to Lawrence E. Spivak, the host of NBC's public-affairs television program Meet the Press, *expressed, with brevity and wit, Bush's unsparing attitude toward forums provided by journalists.*

THOSE WHO TALK FREQUENTLY BECOME IGNORED

Dear Mr. Spivak:

The difficulty is this. Once in a while I speak out when I feel that I have something important to say. But I have found that those who talk frequently become ignored, and that if I speak only at long intervals, I get plenty of attention.

This happened recently in connection with Sputnik where I hammered in a few points and got plenty of attention while doing so. Having said what was on my mind, I will now subside for a while.

Undoubtedly later on, as things develop a bit in Washington, there will be more that ought to be said. But it is impossible to tell just when this hour might occur. This does not fit in with your affairs at all well, as I know, for you cannot shift your program suddenly. But let's see what happens and my guess is that along sometime early in the year there will be a situation that warrants being hammered on as far as I am able to contribute to the discussion.

Cordially yours,

V. Bush

Two years later, in May 1959, Bush agreed to appear as the sole guest on Meet the Press, facing a panel of questioners including Spivak, Edwin Dale of the New York Times, Roscoe Drummond of the New York Herald Tribune and Peter Edson of the Newspaper Enterprise Association. The "unrehearsed" program aired nationally on Sunday mornings on both NBC radio and television, and Bush represented a recently formed foreign-policy advocacy group, the Committee to Strengthen the Frontiers of Freedom, which he chaired. In this exchange, Bush describes the world as divided between free and unfree countries, democratic and totalitarian nations. An advocate of countering Soviet and Chinese expansion through military alliances, Bush gave voice to a "containment" philosophy that was widely shared at the time but would be gravely tested in the 1960s by the Vietnam War. Then as now, many Americans questioned the value of assistance to other countries and some went so far to oppose all such assistance. Bush spoke against these new isolationists in strong terms, arguing that the United States. has a special obligation to assist and protect democratic nations.

Bush's words are selected excerpts from the transcript of the program. The questions, or topics, help put Bush's comments in context but are not always drawn from the program.

On the need for secrecy?

There are some things we don't want to tell the Russians. I think we err in telling too much. There have to be secrets, there has to be classified material. And what do we do? We get the best professional people we can get, through our democratic process, by appointment by an elected president, who have access to all of the facts and to a very considerable extent we have to depend on their professional judgement.

Vannevar Bush to Lawrence E. Spivak, November 5, 1957 (Vannevar Bush Papers, Library of Congress); *Meet the Press* transcript, for May 24, 1959 (Vannevar Bush Papers, MIT Archives). Materials on the Committee to Strengthen the Frontiers of Freedom can be found in the Williams College Archives, Williamstown, Mass.

Is more spending on education necessary for America's survival?
No, I don't think it is necessary to our survival but as an old teacher, I certainly hope we will find some way of raising teachers' salaries in this country.

On why the United States should send other nations weapons made in the United States?
There is a very good reason for that. We might give them the plans and let them make their own, but we can't take the time. The lead time on a complicated missile is several years under the best circumstances. A new model of an airplane takes four or five years before it can be produced. This is a matter of urgency. If we want Europe to be thoroughly armed with modern weapons, as we do for our own defense, as we do so that Russia won't dare to start a war, then the best way we can do that is to see that they get those weapons from us until they can make their own.

On whose views does he present?
The opinions I express are my own personal opinions and if I think the [defense] program that is put together doesn't make sense, I will say so. I hope and trust it will make sense, and I think it will. But merely because we banded ourselves together [as the Committee to Strengthen the Frontiers of Freedom] for the purpose of getting the general idea, the philosophy of the thing, across to the American people as much as we can doesn't mean that we abandon our American citizen's right to criticize.

Are American military forces around the world capable of deterring expansionary thrusts from Russia and China?
There are five million men under arms, among our friends and allies. I think that is a deterrent of some magnitude.

But is that enough in specific spots, such as South Vietnam? Could South Vietnam withstand an attack from Chinese communists?
No, and [South Vietnam] wouldn't have to. But it would certainly make the Russian imperial program pause before they sail into the forces in Vietnam, if they knew that they were one of the great body of free nations banded together for their common defense and bound to come to one another's support it attacked.

On why military alliances with allies should help deter threats from Russia and China?
That is the essence of this whole affair. We do not believe in this country as a Fortress America, of drawing into our shells and waiting for the inevitable. We are not a group of pessimists or defeatists. We believe that if the free nations of this world are banded together in mutual security and mutual defense, defending the entire free world, that that is far better defense than any individual defense on our part could possibly be.

What would happen if the U.S. stopped foreign aid, including military aid, to our allies?
I think it would be a calamity of the first order.

If we reduced our aid, wouldn't our allies do more to strengthen their own military?
No, I do not [think so]. I think they would be thoroughly discouraged at the thought of whether or not we were abandoning this idea of mutual security, and in that discouragement they might not be willing to stand up to the great Russian threat. It is only because they are allied with us, because all the free world is together on this thing that they have the courage to stand up. I think it would be a calamity of the first order for us to put any terminus on this whatever. This plan, this aid, must continue as long as the threat endures.

Why can't Europe stand on its own? Why must we constantly give them money?
One of the best things, one of the wisest things, that this country ever did was when the Marshall Plan [to help Europe recover economically] was put into effect. There isn't any question that it had the hearty endorsement of the great majority of the American people to an extent that I think was magnificent. They understood. They wanted to have it done. They were willing to be taxed to have it done. They believed in it. I believe in the same way that if they [the American people] have a grasp of where we stand now, their enthusiasm and their support will be up to that. . . .

[There is continuing need for military assistance to Europe because of threats from] the Soviet Union plus the Chinese communists that you are talking about. Every aggressor has the grave advantage that he can strike where he will and he can pick his own time.

We on the other hand have to preserve the peace, and that is a far harder thing to do. We have done it thus far and done it very successfully in spite of the fact that Russia has tried to conquer the world, and still says that it would conquer the world.

Remember that before NATO was formed, Russia by one means or another had conquered Poland, Czechoslovakia, Lithuania, Romania—a dozen nations. Since we have banded together under NATO, the progress of Russia in conquering Europe has ceased and stopped. Our object is to stop Russia in its tracks, to prevent war. And it is a far greater task to do that than to be ready to perform an aggression, for an aggressor always has the advantage of surprise.

Is it our European alliance that stopped Russia or our nuclear weapons?
I think personally that right after the war the question that the Russians had in their mind as to how many nuclear weapons we might have in storage had a good deal to do with [stopping Russia]. But beyond that, there are two or three elements in everything of this sort. Of course, the strategic air force has been an important element. So have the ground forces in Europe.

For the war, if it should come, an all-out war, would come in many ways. One would be a sudden, all-out nuclear attack. Against that we must be prepared at all times to so retaliate and to be able to retaliate in spite of a surprise attack, that the Russians could not possibly dare to spring it.

Another way that war might come would be by the Russians starting their armies rolling . . . into some part of Europe. They must be opposed by ground armies. For it is quite possible that that might start and no nuclear weapons be used on either side.

We need to meet every threat.

Would our European allies remain neutral if they faced a possible nuclear attack from Russia?
They wouldn't remain neutral.

You take your two kinds of wars. Suppose Russia suddenly sprung a nuclear attack on the free world. It would be over in a week or two. The question of whether we and our allies could retaliate at once would be the determining question of the outcome. There would be no time for any discussions except to get into the act as fast as possible.

In the case of ground armies trying to roll over Europe, would the European nations defend themselves? Of course they would. They are free people, they are as proud as we are. They have no intention of coming under communist yoke.

How long will the threat last?
The reason that Russia has been able to build an A-bomb, has been able to put a Sputnik in orbit, is because they imitated us. At the end of World War II, I was perfectly sure that the laboratories of Russia, commissar-ridden and ruled by political hacks, could not possibly compete without free institutions. They learned their lesson. Today their laboratories are just as free as ours are and their scientists just as free as ours, and it is because of the giving of that freedom that they are able to perform as they have.

Now you cannot give freedom to a part of a people and not all of a people. You cannot give freedom to a group and hold the rest in slavery. In this way there will be a change in the Russian system. Not for any comfort of ours at the moment, but as a basis for real hope for our children.

What's the role of economic health in the future of the Cold War?
We cannot afford to have Russia allied to totalitarian states all over the world. And there is a very important point there: Communism does not prosper on just the existence of poverty. What determines whether communism can get a hold or not is whether there is progress. If there is opportunity, if the scale of living, however low, is improving, the communists can't do a thing. And a very few dollars compared to our wealth can make the difference in a country between their having hope for the future and having the type of discouragement and poverty that makes them ripe for communist infiltration.

45

ON SPUTNIK (1957)

Bush often spoke as he wrote: plainly, bluntly, and with a moralizer's sensibility about how advances in science and technology make human life more challenging, complicated, and perhaps uncomfortable. In this brief interview with Newsweek *in October 1957, he tried to make sense of Russia's seemingly shocking advance in launching its Sputnik satellite into orbit around the Earth, which aside from indicating Russia's prowess in space exploration also highlighted the possibility of Russian nuclear attack on the U.S. mainland. Despite massive spending on military research, the United States would not match Sputnik for almost four months. While admiring of some Russian achievements, Bush played down the threat to the American mainland of nuclear-armed guided missiles. His failure to grasp the danger was an error in judgement that troubled contemporaries.* Newsweek's *questions are in italics.*

ON SPUTNIK

What was your reaction to Russia's launching of an earth satellite?
If it wakes us up, I'm damn glad the Russians shot their satellite. We are altogether too smug in this country.

Newsweek magazine, October 21, 1957.

There seems to be all sorts of excuses that we could have shot one too, if we had wanted. Does that impress you?
They did it first, didn't they?

What is your estimate of our military research after the satellite and Russia's claim that it has test-fired an intercontinental ballistic missile?
All is not well in our military research. But the really serious thing is that all is not well in our military planning.

How is that?
We do not have unified military planning. The Joint Chiefs of Staff are supposed to do that, but they don't. We have here plans (Army, Navy, Air Force) instead of one. We put great big projects into being at great expertise without anyone reviewing them to determine whether they fit into a unified plan. The program for overseas bases was first proposed by the Air Force, and reviewed by the Air Force. The Navy's carrier program was proposed by the Navy and reviewed by the Navy. I'm not saying these programs are bad, only that they should have been reviewed by some impartial independent body.

What do we need to do to catch up with the Russians?
Unify our military planning. Without it all else is futile, and without it you cannot have unified research.

How do you think we can achieve unified planning?
The only fellow who can do that is the President of the United States.

Do you feel there is any need for a civilian agency to direct the missile program and other military research such as your old wartime OSRD [which Bush directed during World War II while reporting to President Roosevelt]?
No. This country is too prone to try to solve everything by creating another agency or board. The Joint Chiefs of Staff could do it. . . . Right now the Joint Chiefs should be taking a long hard look at the missile program.

What about the missile race? Many people interpret Russia's launching of a satellite as evidence that it is well along toward an operational ICBM [nuclear missile]?

The fact that Russia has shot a satellite doesn't mean that we have the intercontinental missile on our doorstep. We're a heck of a long way from that.

How so?
There are three main problems: Getting the missile up, getting it back into the atmosphere and making it hit the target. If the Russians can guide their missile to the target and have solved the problem of re-entry into the atmosphere, they have done quite a lot, but I don't believe it.

You're aware that many people have appeared to accept at face value the Russian claim that it fired an ICBM over a long range into the target area?
When we know for certain that a missile landed in the target area we will know these problems have been solved.

Dr. Bush, you have said that our immediate need is to achieve unified military planning? What do we need to do over the long pull to improve our science?
We'd better tumble to ourselves, as the Russians have done. Not so long ago the Russians were considered backward in the sciences; now if there's a youngster with a talent for science they make sure he gets all the education he can take. We still have bright boys in this country who can't afford to go to college. There ought to be some kind of program for making sure our boys get all the education they can take.

In Russia if a boy flunks out he is put into the army; in America he is put in the army anyway [via the draft]. This is too serious to adhere to the old principle that all must sacrifice equally in the service of their country. Then, too, in Russia the secondary schools have excellent teachers; the outstanding teacher is a respected man in his community and is paid well. Many secondary schools in this country have teachers who have no science training at all, and we don't pay our teachers enough. In this country a full professor gets two and a half times the income of a laborer; in Russia he gets eight times.

Does all this mean you are pessimistic about our chances of catching up to the Russians?
No, I'm not pessimistic. We have gone through two world wars and survived. I guess we can survive again.

46

"ALL-OUT WAR UNTHINKABLE TO ANY SANE INDIVIDUAL" (1959)

Nuclear war and a Soviet invasion of Western Europe dominated the anxieties of many Americans in the 1950s. In a single paragraph taken from a brief letter to Harry A. Overstreet, a popular writer on psychology and a former chair of the psychology department at City University in New York, Bush presciently identifies the stubborn reality that weapons of mass destruction were a structural feature of everyday life and that nuclear weapons were regrettably only an instance of the wider problem of proliferating weapons of mass destruction. Bush's insight deserves highlighting, and this brief letter is presented in full.

Dear Mr. Overstreet:

I have read your thoughtful letter of 4 June, and I would be very happy to talk things over with you except that I am not in Washington often and usually very busy as you can well imagine.

I think there are several things that people fail to keep in mind when thinking about the subject of your letter. In the first place the Communist land forces are not as numerous on the German border as is commonly believed. They cannot maintain and supply more than a moderate number of divisions on a narrow front. Second, I am sure they [Soviet forces] would hesitate to move without covering all of their satellites with troops

---

Vannevar Bush to *Harry A*. Overstreet, June 9, 1959 (Vannevar Bush Papers, MIT Archives).

for there would undoubtedly be uprisings if they did not do so. So even if there were no nuclear weapons involved, [a Soviet attack] would by no means be a walk-over.

Many people think that if we had no nuclear weapons in the world everything would be much more pleasant. Certainly it would be in several ways. But on the other hand, if there were no nuclear weapons, I feel sure that by now Russia would have developed chemical and probably biological warfare to the point where they would be at least equally terrifying and perhaps even more so. In other words, we do not face a situation involving just one weapon but a general situation where science has advanced to the point where its use in warfare makes all-out war unthinkable to any sane individual.

47

MACHINES TO FREE MEN'S MINDS (1960)

In this short essay for a Boston newspaper, Bush covers the entire sweep of human technological history and imagines new tools in the future for expanding human consciousness and creating a new culture. In vivid and concrete language, he argues that tools define humans as much as humans make their tools. To imagine the future of human tools and technologies, Bush argues, requires a deep understanding of the history of tool making. "Automation is a new word but an old idea," he writes. The long story of humanity, for Bush, centers on the unfolding coevolution of what historian Thomas P. Hughes labeled the "human-built world." Here Bush anticipates a contemporary theme and comes close to identifying and naming a new era in human civilization, which thirteen years later, in 1973, sociologist Daniel Bell described as the "Post-Industrial Society." Bell argued that that postindustrial society would be information-centric and service-oriented and would replace the industrial society as the dominant system. Or as Bush wrote here, in closing, the transformation "means, in short, if we use it to the full, that we can build a new culture" and create "a far greater revolution than was occasioned by man's" harnessing machines "to the aid of his muscles."

MACHINES TO FREE MEN'S MINDS

Much of our prosperity and high standard of living resulted from man's use of complex tools. A part, but a part only, of this evolution is called the industrial revolution. The whole process involved a number of interconnected steps.

The first simple tools merely sharpened men's fingers, or lent weight to his blows. But a great change came when he learned to supplement muscle power with mechanical power, first that of steam and waterpower and later by the gas engine and the electric motor.

A still greater advance occurred when we found that tools could be caused to make more tools. This followed as patience and ingenuity gradually elaborated machine processes and refined their precision. The most far-reaching result involved manufacture by interchangeable parts, and the production line.

No negligible part of the whole development revolved about instruments for measuring precisely, and for indicating often at a distance, the condition of fluid in a pipe, metal in a furnace, or the material flowing through a channel as it became shaped and modified by cutting and molding.

Then, finally appeared machines that could carry out very complex assignments on the basis of a set of instructions, once impressed and thereafter controlling. Automation is a new word but an old idea. Its basis is relegation to the machine of physical or chemical operations which can be precisely defined, and which are repetitive.

With this concept appeared electrical stations, which house and operate power machinery, with no human operators present. Automatic telephone equipment routes calls all over the nation, at the behest of a subscriber who twirls a dial, chooses the best free route, measures the length of [the] call and enters the proper charges in the records for billing. Oil refineries, with miles of pipes, furnaces, pumps [and] storage tanks, go about their intricate processes, with only a man or two watching their performance and occasionally correcting a bit here and there.

A satellite, hundreds of miles above the earth, moving three miles a second, receives a command, takes a picture of the earth beneath, stores it, and on a second command transmits it to a ground station.

Bush's article appeared in the *Boston Sunday Herald*, October 30, 1960.

This whole procedure, of relegating to the machine man's definable tasks, is by no means at an end. It can go forward only so fast as is possible by the reason of skills and savings. The men who design the complicated devices, the men who build the tools which build them, those who keep them in order, diagnosing and correcting their departures from propriety, are men of a new culture.

Automation of a process is indicated only when the saving of labor in its use is greater than the labor involved in constructing and maintaining it, including of course the labor of mining and modifying the materials out of which it is built. Labor here means much more than just man hours, it includes also the skills involved. An automation, where successful, tends to replace those who merely carry out a monotonous repetitive action by those who think and create as they work.

There will always be those who are capable only of a dull repetitive job, and there will always be work for them. But automation can go forward only as there are more men capable of far more demanding tasks. The early stages of the industrial revolution forced millions into lives of monotonous work.

This [next] phase is capable of the exact opposite. It can cause temporary disruption, and unless we take care, local distress. But it can also provide more amply for man's needs and desires, and produce more leisure.

We are now on the verge of another revolution of even more far-reaching consequences. It is creeping up upon us without its presence being fully grasped. It is still in its infancy, but it has already made great changes.

This is the mental or intellectual revolution under which man relates to the machine those parts of his thought processes which are definable and repetitive.

Again, it is no new thing. The first adding machines exemplified it. Automation, in which an automatic machine measures its own product and corrects its own processes as it proceeds, involves the basic idea.

The great digital computing machines now being installed widely are the most powerful devices of the sort thus far developed. They perform only simple arithmetical processes, plus storage and searching, but they do so with extreme rapidity and are little bothered by the complexity of the calculations they are called on to perform. But this is only the beginning.

Someday there will be machines which perform more in accordance with the ways in which the brain itself functions. When we wish to recall a fact or a figure we do not, in our brain processes, consult a numerical or alphabetical index. Neither do we examine every fact in our heads to select the right one. Rather we follow trails of association, stepping from one stored memory to the next, until we come to the desired end. Moreover, we continually add to these trails, modify and refine them.

The attainment of maturity consists largely in the creation of a complex set of threads among our billions of memoirs [and memories]. Someday a machine, under our complete control and orders, will do all these things, far faster than we can do them, and more precisely. Moreover, the memory of the machine will not fade.

All this does not mean that we will be reduced to robots. Quite the contrary. It means that men's minds will be freed form the humdrum drudgery of thought, the evanescent memories, the labor of monotonous repetition.

Freed from these, men's minds can far better tackle the tough thinking that lies in store for us, if we are, for example, ever really to grasp the complex processes of life or the complex interrelations of society. It means also that we may turn to those aspects of existence in which the machine can be only the useful servant, to the subtle and creative, the visionary, the emotional and artistic.

It means, in short, if we use it [machines to free human minds] to the full, that we can build a new culture. This will be a far greater revolution than was occasioned by man's calling the power of machines to the aid of his muscles.

There is here a challenge. For the evolution can come about only as men are trained to meet it, and only if they acquire the courage and pioneering spirit for its consummation. All education has this challenge before us.

48

ON SPACE EXPLORATION

The James Webb Letters (1961–1963)

These letters by Bush to James Webb, leader of NASA and administrative head of the space program, arose from a meeting between the two at the offices of Bush's former outfit, the Carnegie Institution of Washington. Webb asked Bush "if you can give me any advice as to how to proceed" with space exploration. Bush in writing offered wide-ranging advice on approaches to exploring space, while at the same time providing skeptical perspectives based on his view, first expressed privately in 1960 in a letter Bush wrote, that "we in this country are prone to follow fads, and we overemphasize the spectacular." He repeated his claim in his first letter to Webb. In his second letter, he makes a vigorous case for abandoning manned space flight, which he dismisses as "a fairy tale" with scant scientific value. But Bush also concedes he's open to being "regarded as an old fogey" and is nearly "alone" in criticizing the nation's space program.

ON SPACE EXPLORATION:
THE JAMES WEBB LETTERS

May 18, 1961, Bush to James E. Webb

I think . . . in many ways your thoughts and mine on this whole space matter run parallel. We recognize the difficulties of the problem, and we

Vannevar Bush to John H. Heller, January 29, 1960; Bush to James E. Webb, May 18, 1961, and April 11, 1963 (Vannevar Bush Papers, Library of Congress).

know how high the stakes are. I have all the appreciation in the world of the responsibilities you are carrying, and I certainly wish you all sorts of luck and success in the whole affair. The thing that disturbs me personally is the attitude of the American people. I was appalled when the recent reception of Commander [Alan] Shepard [on May 5, 1961, the first American to travel in space, a few weeks after the Soviet astronaut, Yuri Gagarin, did so] took on somewhat the form of a sporting event. The enthusiasm of the American people was in one way gratifying, but in other ways somewhat terrifying as I think it showed all too clearly that they do not realize what a serious affair our contest with the Russians really is. One angle of this appears in connection with the recent series of [labor] strikes and the like on missile bases and establishments. One would think that the American people worried about a missile gap, somewhat terrified at the Russian threat, would rise in its wrath and demand there be no such interruptions in our defenses. But they take it very calmly, and apparently pay no great attention to the matter. This becomes reflected, of course, in our governmental attitude, for we proceed to have some more conferences instead of insisting forcibly that no such group anywhere interfere arbitrarily with our defense program.

Of course you know that I feel that the scientific aspects of space experimentation are being overdone. This does not mean that I think the military aspects are being overstressed by any means, neither does it mean that I do not realize that fine scientific results have been obtained. It merely means that as a nation we are putting too much emphasis on this kind of research compared with other aspects of science which are equally important or more so. We are a nation that pursues fads and which becomes overexcited at times about things that are really trivial in comparison with our truly great problems.

I have felt strongly that the present furor for putting a man in space is the result of a blind mass enthusiasm rather than the result of calm deliberation and comparison. I see the appalling risks of failure in the glare of publicity, and I fear the public does not. I can only hope that our resilience and our freedom of action compensate in some way for our rather naive and immature attitude toward the grim struggle in which we are engaged. And I devoutly hope we take every possible precaution and that we have luck with us.

Democracies are bound to operate in confusion. This is the penalty that we pay for freedom. We have seen plenty of instances in which the

Russian system, to avoid such confusion and going direct to the point, has been able to outpace us. Yet we know that free men can operate more flexibly and more rapidly. Our problem is to keep our free ways from so impeding us that we lose the race, without going over to the other extreme and imitating our adversaries in their methods. Sometimes I get a bit discouraged about it.

The matter of communication satellites is now up for action. Here is a field where the value of results is certainly not in question. But it is also a field in which prior success by Russia would be of special value to them. I trust that nothing will be allowed to impede our speed, and that, for once, we may beat them in a race.

April 11, 1963: Bush to James Webb

I have pondered the subject of this letter for a long time. Now I think I should write it out for you.

Early in the space program, I testified to a Senate committee. As was my duty, I gave my considered judgement, critical of the program.

But since then I have made no public statements. This has been due to a number of reasons. First, I have felt that, being nearly alone in criticism, I would be regarded as an old fogey who could not appreciate the efforts of young men. More important, I hesitated to oppose a program ordered by the President after full advice.

You and I understand this well. During the war I took the position strongly that my job was to transmit to the President the best scientific advice available, and to carry out his orders loyally and without question. I know you have this point of view intensely, for I have seen you respond to the President's wishes many times when it involved hardship or risk on your part.

A part of this attitude has been involved in my relations at M.I.T. There I have taken the point of view that, when duly constituted government called for aid on a program, which aid only M.I.T. because of its unique position could supply, there was a duty to respond, and that my personal estimate of the advisability of the program should not interfere with it doing so.

Now the scene is changing. There are an increasing number of critical editorials and articles. It could change abruptly.

No great program of this sort can proceed without occasional disasters. We have been lucky, and very careful thus far. But some one of these days, a couple of young attractive men are going to be killed, with the eyes of millions upon them. [Such an accident occurred in January 1967 when three astronauts were killed on the launch pad during the first Apollo mission.] Worse, they may be caught in space to die, still talking to us, who are helpless to aid them.

It is often said the public is fickle. It is also said there is, unfortunately, a measure of bull fight complex in the peoples' following of flights. I mean something deeper than either of these. The American public often fails for a long time to utterly grasp a situation and, when it finally does, its reversal of attitude can be sobering or terrifying. A prime example is the Prohibition experiment. A better example is the attitude in 1916. At first unconcerned about a war in Europe, electing a president who would keep us out of war, it suddenly reversed itself and plunged in to halt the Kaiser.

Thus far the public attitude has been one of national pride, enthusiasm over a good show, wonder at the accomplishments of science. It has been uninformed on, or has chosen to ignore, the adverse aspects. It can change its attitude in a month's time. When it does it can be utterly unreasonable, and it can be cruel. I do not know when this will occur; I do not know that it will occur. But I fear it.

Now do not misunderstand me. If I were sure the [space] program was sound I would applaud your driving it forward in spite of any amount of criticism, or any amount of personal risk. And I know you well enough to be sure that is just what you would do. The difficulty is that the program, as it has been built up, is not sound.

The sad fact is that the program is more expensive than the country can now afford; its results, while interesting, are secondary to our national welfare. Moreover, the situation is one on which the President, and the people, cannot possibly have adequate unbiased advice. . . .

While the scientific results of an Apollo program would be real, I do not think that anyone would attempt to justify an expenditure of 40 to 50 billion dollars to obtain them. The [National] Academy [of Sciences] was weak on this point. The justifications given are of quite a different nature. First, it is said we are in a race and our national prestige is at stake. I believe we can disregard the matter of race. I do not know whether there is a race to the moon or not; I doubt it. But national prestige is a far more subtle thing than this. The courageous and well-conceived way the President handled the threat of missiles in Cuba advanced our national prestige far more than a dozen trips to the moon. Having a large number of devoted Americans working unselfishly in undeveloped countries is far more impressive than mere technical excellence. We advance our prestige in many ways, but this was immature in its concept.

I hear that the program will be justified by its byproducts. We might get a billion dollars' worth of benefit that way. I doubt if it would exceed this.

I also hear—and some of my good friends advance this argument in all seriousness—that the program is inspiring the youth of the country, and spurring us on to great accomplishment. It inspires youth all right, and it also misleads them as to what is really worthwhile in scientific effort. In fact, it misleads them as to what science is. It is well to inspire a child, and the use of fairy tales is legitimate as this is done. But when a child becomes a man he should be inspired to judge and choose soundly, to avoid being carried away by mass enthusiasms, to understand the tough world in which he will play his part, technically and economically. It is wrong to inspire him to have an exciting adventure at his neighbor's expense.

I also hear that this is a form of pump priming, that it is a shot in the arm to industry. Anyone who still believes in pump priming should read again about the 1929 debacle, and the sorry following years when we long failed to emerge from the resulting depression.

A most serious point about this whole affair is that the people of this country, and the President with his appalling responsibilities, cannot possibly have adequate sound scientific, engineering and economic advice regarding it. This is due to the very vast size of the project. Nearly every man who could speak with authority on the subject has a conflict of interest. Now do not misinterpret this to mean that the scientists of the country are all feeding at the trough, and so selfish they would subordinate their judgement as to what is true to what is advantageous to them. There are some of these of course. I even hear rumors of artificial pressures being brought to bear on individuals and companies to ensure their conformity, but such rumors always float about when there are great undertakings, and in any very large organization there are always subordinates of little sense, as we have seen exemplified often.

I do not mean this sort of thing at all. The scientist or engineer in a university or a company is in a difficult quandary. He may honestly believe the program as a whole is highly fallacious. But it has been decided upon at the top level of government. It is supported by his colleagues, many of whom have enthusiasm. His organization has been urged to participate. Who is he to stand out against this powerful trend? He consoles himself by Cromwell's admonition, "I beseech you, bethink you that may be mistaken," and sides along with the crowd.

We pride ourselves that, in this democracy, the minority has the opportunity to speak. Yet it takes courage, and an unusual sort of detachment, to stand against a nearly unanimous opinion of friends and colleagues, and to risk one's reputation in a futile attempt to halt an avalanche. I know this whole program has never been evaluated objectively by an adequately informed and disinterested group, and I fear it never will be.

The whole problem is in the hands of the President, and he has many problems on his mind today. He leans on you to steer him straight. As we now go, there is danger ahead for the program and danger to his prestige. I hope he will alter his handling of this whole affair before a balky Congress, or public opinion, forces him to do so.

You and I think alike on the tough problem of the relation to the President of a man on his team; we have discussed it a number of times. Your creed and mine depends on two main principles. First, the President on a problem should have the best advice this country can afford, with differences of opinion where there are any faithfully transmitted, and it is the job of the man who reports to him in an area to see that he gets it. Second, when the President, with full grasp of a subject and thus advised, makes a decision and issues an order, it is the job of his lieutenant on the subject to carry it out loyally and effectively whether or not he agrees with it. This is especially true in time of war, but it is also true of a key subject in time of cold war. The only exception would be a situation in which the lieutenant's disagreement was so complete that he found himself unable to perform well, in which case he should step aside and incidentally say nothing.

I believe the President could alter his attitude and his orders without a reversal of form which would embarrass him.

I know what I think should be done. As a part of lowering taxes and putting our national financial affairs in order, we should have the sense to cut back severely on our rate of expenditure on space. As a corollary, he could remove all the dates from plans for a trip to the moon; in fact he could announce that no dates will be set, and no decision made to go to the moon, until many preliminary experiments and analyses have rendered the situation far more clear than it is today. He could lop off, without regret, marginal programs that cannot be soundly supported, and continue only where results are clearly attainable and worthwhile, in weather and communication satellites for example. He could order experimentation concerned with long space flights confined to those features which are clearly central and determining, avoiding hardware except where it is necessary. Then, after a year or so, the entire program could be reviewed through a professional disinterested board, made up of scientists, engineers, economists, financial men, and men with keen judgement of public attitudes here and abroad.

By so doing, he could reduce the rate of current expenditure at a time when any such cutback would help him in his tax program. He could avoid commitment to vast expenditures until such time as economic prosperity justified them, and through analysis had shown them to be warranted. And I believe he could do this without real damage to an overall logical sound space program.

There were times when you and I both reported to the President, and we worked closely together in doing so, even when we did not totally agree. Today you are still doing so, while I have dropped out of the active picture.

But, whatever you do, and however the program may work out, you have my best wishes, my deep personal regard.

49

THE OTHER FELLOWS' BALL PARK (1961)

In this brief reflection on the politics of science, Bush presents an image from ordinary American life to convey complex ideas and nuanced distinctions. While Bush himself displayed no special interest in competitive sports, either as a participant or a spectator, he surely knew well that Americans are a sporting people. In this image of the home-team playing on the road, he has in mind rival baseball stadiums because, alone among professional sports, baseball fields are asymmetric. While every pro football and basketball field has the exact dimensions and "ground rules" dictated by their respective leagues, only baseball permits each team to have a field with distinct dimensions. Bush liked to view science, engineering, politics and the military as each possessing diverse fields of play, governed by somewhat different rules. From this starting point, Bush needs to take only a short step to the idea of self-governing domains, akin to "republics," that define their terms and agendas, and impose distinct norms and rules on members. In this way, science and politics, like baseball, can be viewed as a contest between rival tribes who seek to advance different but sometimes complementary programs. While the sporting imagery suited Bush in an era of "big science," large government programs and a relatively unified national agenda, today the image may seem anachronistic if not outmoded. But at the time, science leaders widely circulated Bush's perspective, which he originally shared in a speech to research administrators.

THE OTHER FELLOWS' BALL PARK

There is an old saying that when one goes to play ball in the other fellow's park, it is incumbent on him to learn the other's ground rules. There is no use protesting about these rules or asking that they be altered. The reasonable course is to find out what they are and proceed accordingly.

In the same way the scientist entering into the affairs of government, and into the political arena, has the duty—if he is to be fully effective—to find out how men in this arena actually function. It is not sufficient for him to criticize their methods, and it is fatal if he takes the point of view that he is now in a section of society which is governed by rules less ethical or less advanced than his own. We live under a democratic system. An essential feature of this system is the means by which men acquire and maintain political position and authority. These means involve a thorough understanding of human nature and of mass reactions. Many men in political life are masters of this art. And in the large majority they are also devoted to the welfare of their country. As a result we have effective working of the democratic system as far as efficiency in government is possible. The democratic system creaks at its joints, it wastes time and money, it bases its decisions more on subtle influences than on rational logic, and it often irritates those who are accustomed to more orderly and systematized functioning. But it just happens that the democratic system with all its faults is the best system of government ever devised by the mind of man. Scientists today are privileged to participate in this whole affair to an extent never before true in this country, and it is certainly incumbent on them to understand and indeed to sympathize with the local ground rules which govern the ball park in which they are now exercising important influence.

I have seen great damage done to the whole scientific community by the eminent scientist who, appearing before a congressional committee, made evident his general contempt for the individuals before whom he appeared, and who talked to them as he would to a group of schoolboys. Fortunately in the years since the [second world] war, scientists have matured in this regard and this does not now occur.

Excerpt from an address by Bush to the National Congress on the Administration of Research, San Juan, Puerto Rico, October 10, 1961; reprinted in *Science* magazine, October 20, 1961.

We need to go beyond this in our thinking. As scientists and engineers we nevertheless regard with admiration and respect the subtle functioning of a medical man in a difficult case, not relying upon the science which underlies his art, but relying upon the art itself, who rescues a patient from an obscure source of distress. We admire and respect also the artist who, knowing very little about the physics of color or of light reflections, produces a work of art that stirs emotions or revives long-forgotten memories. In the same way we need to learn to respect, in fact to admire, those individuals who are masters of the art of operating in the confused arena of the American political scene, especially when this subtle undefinable skill is joined with a broad altruism. In fact, if scientists are to have their full influence for the good of the country in the days to come, many of them will indeed need to learn to practice this difficult art.

50

TWO CULTURES (1962)

Bush was ever alert to the twin pressures of over-specialization and the gulf, or alienation, between those working in the arts and humanities and those working in science, engineering, and business. In this text of a speech, he advanced an optimistic view that literary and artistic people can and should maintain a basic understanding of rapid changes in science and engineering. In thoughts shared with a group of alumni of MIT, Bush examines an influential argument by the British scientist C. P. Snow about the costs and consequences of the estrangement between the "two cultures" of science and the humanities. Snow bemoans the estrangement between the scientific and technological and the literary and artistic; and he fears that the gulf between the two cultures may be insurmountable. While privately sneering that Snow "wrote about two cultures simply for the sake of trying to sell his books," Bush recognizes the problem and argues that "the division . . . is arbitrary and makes no sense." He contends that in modern times both breadth and depth are possible for the "cultured" person, and that only a kind of snobbery makes the literary and artistic appear more "noble and elevating to the spirit" than scientific and engineering activities. Bush even insists that there are more than two intellectual cultures, or camps, and perhaps many more. While admitting "the time is past when great minds could encompass substantially all knowledge," Bush says that cultured people should "acquire genuine breadth without at the same time losing the ability to excel in some intellectual specialization." He goes on

*to ask a question that remains relevant today: "What, then, should be the
nature of education either formal or self-conducted which will inculcate
this type of culture?"*

TWO CULTURES

It has recently and often been asserted that there are two cultures, and
that these cultures are to a considerable extent mutually exclusive, and
bound to be more so. The first kind is asserted to be scientific, involving
an understanding of things, and the second liberal, involving an understanding of men, their history and their emotions. Sometimes it is implied
that the former is crass, narrow and painfully utilitarian, while the second is noble and elevating to the spirit. I wish to disagree with this whole
absurd bag of tricks.

Certainly there are two kinds of culture, and certainly it is worthwhile
to distinguish them and to develop them intensively and with discrimina-
tion. But they do not divide up in this way.

Any man who understands things, and is not interested in people, is not
cultured. He has not been educated, and has not educated himself. Also,
in this modern world, when the applications of science are transforming
civilization, any man who is deep in history or literature, and who shirks
from an elementary understanding of the great evolution in science, while
he may be polished, is nevertheless grossly ignorant. This does not mean
that a single individual can qualify as an expert in a dozen such diverse
lines as atomistics, oriental pottery, solid-state physics, medieval poetry,
electronics and Latin. The time is past when great minds could encompass
substantially all of knowledge—if it ever existed. It is rather a matter of
interest, and a determination, to acquire genuine breadth, without at the
same time losing the ability to excel in some intellectual specialization.

The division into scientific and liberal is arbitrary, and makes no sense.
Moreover, a conviction that such a division is desirable or inevitable can

---

"Two Cultures" is the text of a speech Bush (then honorary chairman of the MIT Corporation) gave
to members of the university's alumni association in 1962; reference to Snow "trying to sell" books,
page 319 of Bush's oral memoir, dated 1964–65, Carnegie Institution of Washington.

do harm; it has already done so. But I assert there are two varieties of culture, and that a grasp of their characteristics is worthy of thought. To be clear, I should define. And I would define the first type of culture very simply: it is the basis of wisdom. It involves an understanding both of nature and of men. It provides the foundation upon which a wise man may base judgements concerning everyday affairs or great issues. It provides the background for the conduct of a useful and satisfying life, and for a salutary influence upon the lives of others.

What, then, should be the nature of education either formal or self-conducted which will inculcate this type of culture? This can best be answered by asserting what it should not be. First, it should not be over-specialized to the exclusion of breadth. The man who studies the chemical aspects of genetics, who becomes a master in that field, and who has no interest whatever in the current political scene, or the masterly eloquence of a Churchill, or the psychological insight of Shakespeare, is not cultured in accord with our definition. Neither is the individual who develops a reputation as an archaeologist of the Mayan civilization but who cares not a whit for the influence of the transistor on modern communications. Second, it should not be superficial. The chap who can talk impressively, for three minutes, upon any subject under the sun, from moon probes to Cretan water works, but who cannot talk intelligently for 15 minutes, among experts, on any subject whatever, is just a dilettante. He may have breadth but he has no depth. Wisdom requires far more than extensive superficial knowledge; it requires also the ability to reason, whether by strict logic or by balance of evidence amid contradictions, an ability which is attained only by thinking intensively and exhaustively, surrounded by keen minds doing the same thing, competitors struggling for mastery, in a tough demanding subject.

It does not matter too much what that subject may be, provided it is joined with a breadth of interest in adjacent subjects and more widely. Our cultured man may be a lawyer or a surgeon or a business executive or a scientist or a banker or a linguist. There is no area of man's accomplishments which does not require, for eminence, discipline in the development of thought processes. The mathematician and the skilled clinician use widely different methods of reasoning in their work, but both types are needed for the effective handling of affairs in our complex civilization. The essence of acquiring a solidly based culture is that it should simultaneously create a trained mind capable of accomplishments in a specific area, and a broad range of interests effectively pursued by reason of that acquired skill, to the end that judgement may be soundly supported, to the end that there may be wisdom.

It should be emphasized that this sort of culture is not merely utilitarian. True, most of us earn our income by use of our special skills. True, the country needs a wide variety of specialists, highly capable in their specialties, if we are to manage our intricate social and industrial relations to our advantage. True, also, we live in a hazardous world, and our safety and material progress are intimately interconnected. But a culture could support all these things and still be inadequate. There is more to life than material or professional accomplishment. Wisdom which extends only over the area of self-advancement is not wisdom at all. Success, measured in terms of the accumulation of a fortune, or even in terms of public acclaim, is still cramped into narrow confines. To live a full life man needs to reach much farther than this; his primary motivations of self-preservation, the urge to perpetuate his genic constitution, the desire to accumulate or conquer, these are the beginnings only of a truly civilized existence. Beyond these is a sublimated deep yearning to play an effective part in the great experiment on which mankind is embarked, to be a real factor in the evolving fabric of a civilized life. It is exemplified in a practical idealism, in seasoned altruism, in a genuine interest in the welfare of fellow men. Like all primary motivations it leads at times to perversions, and to eccentricities. But only when the goal is thus broad, even though it be but dimly perceived, is there true satisfaction in living. A culture, to be genuine, thus extends far beyond the basis for acquisition, far beyond even the basis for effective contribution to the material and professional affairs of the times; it stretches toward something much higher and much deeper. It aims to provide the basis for true wisdom. And only by pursuing this shadowy goal is there deep and abiding satisfaction in living.

———— ⁂ ————

There is a common fallacy in viewing this whole subject. We speak of necessity of experts, of professional skills, of adding to man's knowledge of his environment and of his relations with his fellows. From this we are apt to conceive of some sort of cultural aristocracy, some exclusive class to which only a few may belong. True culture knows no such boundaries; wisdom is not confined to manipulation of the intricate. Skill, which aids our great objectives, is admirable wherever found. A democratic society in which wisdom appeared only among a chosen elite would soon wither.

Consider a skilled mechanic, one who can fabricate the parts of a gyroscope inertial guidance system, working to tolerances far less than the wavelength of light. He is an accomplished man, contributing uniquely to our affairs. His skills have been built on assiduous study and practice over many years. The mysteries with which he deals are as fully mysteries to the great doctor or lawyer, as theirs are to him. He belongs to no national societies, writes no degrees after his name, makes no speeches. Yet he may be far more an accomplished citizen than his boss, or his boss's boss. Among the multitudes of such individuals are also those who have developed a broad range of interests, and a sound grasp of events and affairs. They are men of influence in their own circles. Without them our political structure would falter. We give them scant recognition and secondary material rewards only. But we would do well to avoid all snobbishness when we think of culture. These are cultured men. They are also, often, wise men. They often lead lives of more genuine satisfaction than those we acclaim. True culture must extend through the entire fabric of society, if our experiment in self-government is to work. Education, well-conceived and carried out, can help enormously, of course, in creating it. But the essential criterion for its furtherance is general recognition that it is worth the effort necessary to achieve it, at whatever level of social stratification it may appear.

Having defined one type of culture as the true basis of wisdom, have we exhausted the field? By no means. There is a second type, by no means secondary. To live wisely in the complex affairs of the day, and with the influence of future trends, is salutary, and extends far beyond mere material success. It can yield true satisfaction. But even this is not to live fully. There is a second type of culture which transcends even this. It cannot be taught, although it can be exemplified and inspired. It yields no concrete definition.

You all know what I mean, although I cannot tell you! The honking of a wedge of geese on a still night. The pirouetting of a lovely child as she practiced her fancy steps alone. The sparking of water in a turbulent brook. The dimple on a still backwater as a fish rises. The sighing of wind in the pines. The subtle nuances of great music. The majesty of soaring columns in a cathedral. Who can classify or define when such things appear, who would wish to? Moreover, who would degrade by trying to tie their appeal to the emotions into some system of cerebral evolution? Yes who would say that any man is truly cultured who is blind to all this?

We know that appreciation of many such interests, which stir the emotions deeply, can be enhanced if a technical basis is laid for better grasp and understanding. The appeal of the graceful soaring of a great bird is still an emotional appeal if one understands a bit of aerodynamics. A thrill from viewing a great painting is more intense if one knows something of the long history through which this art has evolved. A modern symphony can be enjoyed only by those who have lived closely with music for many years. But it is the techniques only which can be taught.

This second type of culture has nothing whatever to do with utility. It has little to do with wisdom in terms of men's current affairs and material aspirations. But it has much to do with wisdom in a deeper and wider sense. Much of it is inborn. Yet it can be developed and enhanced. And it can yield a joy in living which is far more genuine than that which comes from just being an effective cog in an intricate scheme of material and social existence.

One more thing needs to be said about it. Men grow old, and skills fade. A life of professional accomplishment ends, long before life itself terminates. The gray hairs come, the joints creak, the eyes grow dim and the young men take over. But the joys of this second type of culture never fade.

For our safety and progress, for the success of our political organization, for our material welfare and for our physical and mental health, for all that is implied by successful and progressive civilization, we need to enhance by all means possible, and in all strata of society, the culture which is at the basis of wisdom, wisdom regarding the control of nature and the affairs of men. But, in doing so, we should never forget that there is beyond this a culture, and a wisdom, undefinable, intangible, which is essential to a full life, and which may indeed, at some day in the long future, be far more meaningful than merely learning to manage nature, or to order the relationships among men.

There are thus two cultures, one of which can be defined and purposefully developed, and the other only vaguely grasped. But this is a far cry from trying to separate the interests of men arbitrarily into the scientific and the liberal. Our first culture, on which we depend for our well-being and our intellectual advancement, cannot be thus fragmented without destroying its inherent nature.

Now why do I struggle to expound this subtle matter before the group which is gathered here today? . . . We gather because we enjoy the association with successful fellows in a common cause. We take justifiable pride in being part of a great and successful undertaking. We remember all that [MIT] did for us in our younger days, and we are happy to work and to contribute our efforts with confidence that it will still more effectively instruct and inspire the generation which follows us. We band together to the end that worthwhile youngsters may have a heightened opportunity to serve society usefully, and to live lives of accomplishment and satisfaction. We have a common interest because we know that this institution leads in the essential task of rendering our country safe and prosperous in the hazardous days that lie ahead.

But beyond this there is another reason why we meet in common endeavor. In the confused search for the basis of a true culture, we believe this institution of ours is in the forefront. Leading in science and engineering, blazing new paths in a better understanding and control of the physical world, [MIT] is also leading in a sound broadening of education to cover also the understanding of man's social and political interrelationships, his history, his psychological intricacies, his gropings toward law and order, his struggles to govern in an increasingly complex world without inhibiting that personal freedom and opportunity which has rendered this nation great and prosperous. Unhampered by hoary traditions or vested interests, it seeks to create a new and more genuine basis of true culture. We may differ on how this can best be accomplished, but we join in the conviction that the effort to accomplish it is well worthwhile. We join with enthusiasm because we believe cultural background will form for him a true basis for wisdom. We strive because we feel that here he will be inspired to go forward to a life of accomplishment and satisfaction, and beyond this, to a life of joy in living.

51

AUTOMATION'S AWKWARD AGE (1962)

Bush's concern about possible job losses from technological innova-
tion seems contemporary. He cared deeply about making technologies
"humane" and consistent with human values, and he painfully recog-
nized that advances in science and technology often outpaced the capac-
ity of individuals and societies to adapt and strike a balance between
the old and the new. There would be losers as well as winners. While he
never invoked William Ogburn's cultural lag thesis, which the Ameri-
can sociologist popularized in the 1920s as an explanation for the desta-
bilizing effects of technological change, Bush implicitly believed that
humans would "catch up" over time to technological innovations. While
he accepted that some innovations would challenge, or inevitably out-
strip, the capacity of humans to manage and cope with change, he also
believed that humans could take pride in what he called "our present . . .
age of reliable complexity."

AUTOMATION'S AWKWARD AGE

When the Industrial Revolution really struck England, the land of its birth,
in the early years of the 19th century, the outcry against it was enormous,
and no phrase of protest was more memorable than "the dark Satanic

mills" against which William Blake so eloquently raised his poet's voice. The Factory was the locus of all Evil: it represented poverty, exploitation and misery. The cotton mills of Manchester stood as a world-wide symbol of Paradise Lost.

It took a 20th-century English philosopher, Alfred North Whitehead, to point out that most of the terrible sufferings which accompanied the Industrial Revolution were in fact unnecessary and were not caused by the steam engine or the power loom; they were instead the consequences of a giant step achieved in technical innovation with no corresponding step at all in social, political or economic innovation.

Within little more than a century we have hauled about on a completely new tack: whereas in the 1840s, the very existence of the factory system was deplored, by the 1960s the factory has become so [embedded] in our social thoughts that a decline in factory employment is now being widely heralded as the New Doom. Yet the heralded cause of the New Doom is the same as the cause of the Old Doom (1789–1840). Only its name has changed.

There is nothing new about automation except the name itself. It has been going on for generations. The automatic loom, power feeds on lathes, even the early scheme of making a pumping engine operate its own valves, were all forms of automation. In recent years the process has simply accelerated, for two reasons: One is the great increase in wages, which has put a premium on saving labor. The other is the advent of new devices, which lend themselves to be used in complex automatic machinery, and which are reliable.

This last point is important. If we have an automatic assembly including a thousand small devices: relays, valves, condensers, thermocouples, and so on; and if each of these is likely to fail even once a year, it is clear that the whole array will almost never be workable, but will be under almost continual repair. It was only when such things became ultra-reliable that great automatic systems, such as the automatic telephone system, became practically possible. We could well call our present time the age of reliable complexity.

The thought is abroad that the inevitably increasing use of sensing mechanisms, tape-instructed machines, more and more sophisticated

Originally published in a newsletter circulated by Elmer Roper & Associates, a public-opinion survey company, and reprinted in the *Saturday Review*, August 11, 1962.

feedback circuits, is going to put almost everyone out of work and that the mass America will then stand about with its mouth open and nothing to do: a nation of village idiots watching a mechanized parade go by. A more nonsensical notion transcends the imagination. There will always be enough work to do in America, and everywhere else, to keep every man and woman busy for as many hours a week as future societies think desirable. A glance of the nation's slum-ridden cities, or its overcrowded schools, or even a glance at your own backyard—or mine—will verify this simple statement.

It is of course true that automatic devices can throw men out of work, locally and temporarily, if no steps are taken to shift men to other jobs and, if necessary, train them for them. The automobile industry is a good example. It would be quite impossible, in anything like its present form and extent, without extensive automation. When the auto became popular it knocked out trades like carriage building and harness making. But it created 100 jobs for everyone it made obsolete. And it created wealth, widespread and real. The man who owns a modern car, even a compact, is decidedly wealthy in terms of the true value of his personal possessions, compared with the tycoon of a few generations ago driving a horse and buggy.

People forget that automation saves labor in production, but calls for labor to build and operate the machines. If the latter is more costly than the former, it does not make sense to have an automatic machine at all. But every good job of automation decreases the cost of production: if it does not, then it is not a good job. Decreasing the cost inevitably decreases the price of the product in any field where there is real competition. And this means more sales. More sales means more jobs.

The other main objection raised against automation has been that it makes for the creation of monotonous jobs. The usual picture is of a man sitting by a belt on which washing machines go by, and drilling just one hole in a certain place in each one. The absurdity, of course, is that this is just the sort of task which is made automatic. Unless the company making washing machines were dumber than most companies, the man would soon be replaced by a little gadget to drill the hole. If it were too dumb to make this change, it would soon be taken over by a more intelligent company. And the man, if he were reasonably adaptable and alert, would soon be doing a job that is far from monotonous: building such devices, or adjusting and maintaining them.

Actually there is a limit, in fact there are several limits, to the amount of automation that can justifiably be carried out. No automation is worthwhile unless it will reduce the overall cost of production. This includes the important element that it must be reliable enough so that costs will not be unduly increased by down time (a period of machine inactivity). In estimating this, one has to remember that down time of one group of machines may stop production on many others. But from a broader basis, from the standpoint of the country as a whole, automation can proceed only as fast as there are skilled men to carry it out. If by some magic, a machine shop a century ago had acquired a modern automatic screw machine, it would have stood idle, or soon become wrecked, because there simply were no men about with the understanding and skill at mechanical gadgetry to set it up properly to do the job, and this included the boss of the shop and its owner. . . .

I am far from saying that we shall not encounter serious problems as mechanization proceeds at an exponential rate. On the contrary, I am sure we will. The lot of the poorly educated or insufficiently skilled, always hard, will become harder. The man-in-the-rut, who loves his rut, will need retraining at society's expense and, as we have already come to now, the retraining will not always take. The old, like the young, will need more care, shelter and protection. All these groups and some others will need greater help from society than they are at present getting. This is the price of progress: it need not be too high for what is thereby achieved.

52

WHAT IS RESEARCH? (1963)

On November 21, 1963, the day before the assassination of John F. Kennedy, Bush delivered prepared remarks to a Congressional committee on government research. In introducing a list of principles for understanding scientific and engineering research, Bush warned that government-funded research, while essential, "has been overextended." He told lawmakers, "I believe this is an opportune time for a thorough study by this committee. I trust, if your study indicates that we should now apply the brakes [on spending], we will not commit the converse error, and cut back really worthwhile efforts." To help think about what counts as research worthy of public funding, Bush described with clarity and concision twenty-one basic principles that inform wise decisions about this often-misunderstood human activity. His principles illuminate enduring questions about what research is, who does it, how, why, and to what end.

WHAT IS RESEARCH?

Gentlemen:

Your committee has before it a very heavy task, and I am happy that it is being undertaken. It is time for a broad and intensive examination into government support of research.

After the war, and as a result of the success of our scientific programs in developing new weapons, this country plunged into a broad program of government support of research. A report to President Roosevelt, recommending such a course of action, summarizes the opinions of scientists at the time, and is worth review. It is well that this occurred, for the necessary size of the country's research program, to buttress our military strength, to ensure rapid economic progress, and to give us our proper stature in the creation of new knowledge, was far too great to be supported by private funds. Those of us who recommended the program recognized the dangers but felt they could be avoided. In general they have been and the program has gone well and made sense. But the American people seldom to do things moderately. The program has been over-extended and it is still rapidly growing. I believe this is an opportune time for a thorough study by this committee. I trust that, if your study indicates that we should apply the brakes, we will not commit the converse error, and cut back really worthwhile efforts.

In the time available I will not be able to present fully any one aspect of your problem. I will hence confine myself to a number of brief statements of points I feel you should have in mind as you proceed. Some of my statements will no doubt be objected to vigorously. I could add many more but have tried to pick out important aspects of your overall program.

1. What is research? It is the search for new knowledge. But this includes a mathematician with a pencil and paper, sitting under a tree, and it also includes a hundred men designing a new jet aircraft.

2. Much of what we call scientific research is really engineering research, or even straight engineering. Putting a communications satellite into orbit, and working, is an engineering job, incidentally a magnificent one. I trust that when I say scientific research you will realize that I often mean to include engineering research.

3. A man sitting at a desk and thinking is not an expensive proposition. A scientist directing a team and operating an expensive array of apparatus is. The costs of research go up very rapidly when one gets into hardware. When money comes easily there is a tendency to rush into

Statement by Vannevar Bush, Select Committee on Government Research, November 21, 1963 (Vannevar Bush Papers, MIT Archives).

use of complex equipment—too fast and too far. We [as a nation] may be making this mistake.

4. If the country pours enough money into research, it will inevitably support the trivial and the mediocre. The supply of scientific manpower is not unlimited.

5. Duplication in research can involve waste. But on really important problems duplication is inevitable, is even necessary for rapid progress. Competition among scientific groups is as important for producing outstanding researchers as competition among football teams is in producing greater quarterbacks.

6. As has been said many times, you cannot have great advances in applied research unless you have as a basis an extensive body of fundamental knowledge, developed over years by basic research. The development of an atomic bomb was possible only because of fifty years of basic work on the properties of the atom. We once leaned on Europe for our basic research, and Americans turned more naturally to applications. We are now doing better. But we should lead in every important field of fundamental scientific knowledge. Many of these will be hard to understand by layman, or even scientists in other fields.

7. It is impossible to calculate the efficiency of a research program. Having directed all three, I would, for effectiveness in basic research, rate universities and research institutes first, commercial laboratories second, government laboratories third. One reason is that a scientist in a university can afford to make mistakes. Of course there are exceptions: outstanding basic scientific accomplishments [occur] in commercial and government laboratories. But the normal home of basic research is the university.

8. When scientific programs are judged by popular acclaim, we inevitably have overemphasis on the spectacular. That is just what we have today. The deeply important scientific advances moving today are not easy to understand. If they were, they would have been accomplished long ago. Outstanding scientific progress, which will most affect the lives and health of our children is not grasped by many. For example, I rate the programs on interferon highly. Yet, if I tackled a hundred citizens at random, I probably would not find one that knew what it is all about. I might even find the same situation among members of

Congress. Yet it will do much more for us, a decade from now, than a probe of Venus.

9. In any broad program of research the key word in regard to any one aspect of the program is relevance. It is a good word to have in mind in examining any research program. Competent directors of research know what it means. Probably "conducive to progress toward the main object of the program" is as good a definition as any. Just finding out something new is not by itself sufficient justification for research. It needs to mean something when we find it.

10. It makes sense to ask a young researcher in basic research what he is trying to find out, what sort of knowledge he hopes to have at the end of his program which does not now exist. Surprisingly often the answer will be hard to extract. But it makes no sense to ask him just how he is going to do it, what it will cost, or how long it will take.

11. In the nature of things the application of the physical sciences produces interesting gadgets of great popular appeal. The application of chemical science produces a new liquid in a bottle, or a stronger material, or a lowered cost of manufacture, and these are seldom spectacular. The application of biological science produces new drugs to cure our ills, or ingenious ways of combatting insects without injuring birds. These may be very important indeed, but they involve little of the gadgetry which the American public enjoys. We need to use care that our American love of gadgetry does not lead us astray in our emphasis.

12. The salaries of scientific researchers has gone up rapidly in recent years. This is good. They are still going up rapidly, which is also good. But one wonders what will set the ceiling. The dollar seeking a profit, the dollar from [a charitable] endowment, have a tough time competing with the tax dollar. With many agencies using tax dollars and competing for men, we are headed for trouble.

13. We are not doing too well in our rate of industrial growth. We have trouble on flow of gold, and our ability to compete and to export is part of this picture. It would be well to find out whether the pressure of government-supported research is not now preventing industry from doing its research job well.

14. The spectacular success of applied research during the war led to a fallacy entertained by many. It is that any problem can be solved by

gathering enough scientists and giving them enough money. To solve the problem of the common cold, assemble a great institution, fill it with scientists and money and soon we will have no more colds! It is folly to thus proceed. The great scientific steps forward originate in the minds of gifted scientists, not in the minds of promoters. The best way to proceed is to be sure that really inspired scientists have what they need to work with, and leave them alone.

15. Since the war we have avoided almost completely one of the great dangers of government support of research in universities, namely dictation to the universities on their programs. This has been due to a number of things; wisdom on the part of government officials for one. Much of it is due to the fact that judgement of scientific progress has usually been made by competent committees rather than by individuals. This too has its dangers, but it is by far the best way.

16. It is a strange thing to have academic basic research supported by the military services. In my opinion they have handled it very well thus far. But that does not prevent it from being unnatural.

17. I have said nothing about the social sciences, for one reason—because I am not a social scientist. You gentleman are in a better position to judge here than I am. You are at least practitioners in one field of applied social science, or you would not be here.

18. I have also not said much about the space program. I think my [critical] attitude on that is pretty well known. And I doubt if the committee wants to plunge into this question now.

19. I have been speaking about non-classified research, and especially about basic research. There is also the enormous field of military research. I have great sympathy for this committee in their task of judging our programs in the open research field. I suspect that sound judgement on military research is just as badly needed and almost as impossible to arrive at without far more effort than I believe is now contemplated.

20. Since the war we have seen a strange, and to my mind, dangerous development. The Armed Services have called upon universities to manage great programs of research and development, involving secrecy, and often calling for business judgement. Some of this has been avoided by the creation of independent non-profit organizations. We ought to find a better way. The universities will respond, when called upon

by the government to undertake burdens in the public interest. But management of secret programs is not their proper business, and they should not be thus utilized. We ought to be ingenious enough to avoid loading our universities with tasks which may interfere with their proper function of turning out educated men and women.

21. It should never be forgotten that the main task of the universities is to educate people. The country will need skilled professionals in the future as much as it will need new knowledge. As we now go we are not meeting this challenge sufficiently. Every research program placed in a university should be so ordered that its product is not only new knowledge but skilled educated people.

I have suggested a broad range of inquiry. You have my sympathy as you delve into it.

53

THE ART OF MANAGEMENT (1967)

Bush emphasized the role of organization in the innovation process. While he appreciated the importance of inspiration and creative leaps, he believed innovators neglected management at their peril. In this regard, Bush anticipates, by a full half-century, what has become the hallmark of techno-science, both inside the university, within government research laboratories and inside the private technology corporation. Management and organization, perhaps more than any other controllable factors, shape the outcomes of research and development. During World War II, Bush learned to manage innovation at a scale, through large organizations and masses of scientists and technologists, close to production. Bush thought deeply about diverse styles of management and their effect on the knowledge enterprise. His singular contribution to the Manhattan Project, for instance, was his insistence that to achieve breakthroughs required vast resources and the support of large, complex organizations that, despite their size, supported small teams and even solitary researchers. For Bush, management and leadership overlapped and, perhaps intentionally, he didn't sufficiently distinguish between the two. A longer, and less effective, version of this essay appeared, under the same title, in Bush's collection of essays, Science Is Not Enough. *Bush consistently argued that success in science and engineering depends on strong management and organization; for him, knowledge alone was never enough.*

THE ART OF MANAGEMENT

All my life, which has now lasted something under 100 years, I have observed managers in action, in government, in universities, and in industry. And I have puzzled about what made them tick. I have seen poor managers and good managers. The poor ones are fairly easy to catalog. Some of these labor under an inferiority complex and compensate by being martinets. Some seem to have a Messiah complex and endeavor to regulate the private lives of all their staff. Some of course are just plain ignorant. It is much harder to state just what it is that makes a manager really superb. I have had the privilege of knowing many who thus excelled. My attempt this evening is to probe a bit toward the essence of their success. No man can ever fully understand another, even a close friend. But I think a few comments may help to get at the heart of the matter, what indeed characterizes the great manager, whom we all respect. If you will call me back, say in another 25 years, I will then try to give you my observations in a more explicit form.

Our free enterprise system is rapidly being transformed by the advent of the profession of business management.

This new profession differs from the older ones in many ways, and resembles them in some. It does not have the mysteries and trappings of the Old Guilds, no Hippocratic Oath, no admission to the Bar, no system of indoctrinating neophytes, no rigid path of entrance. [The new business profession] resembles [older professions] in that it assumes its practitioners have special skills and knowledge which they will exercise with dignity for the public good. Like the old professions it has its charlatans, its nincompoops supported by special privileges, [and] its snobs. Like them too its protection against these parasites lies in the power of common opinion within its own ranks.

Being young and not yet frozen in its ways, [the profession of business management] specifies no discrete scope of training or qualification to enter its ranks. Many paths, through finance, production, engineering [and] marketing, lead to the top of the profession. Increasingly its ranks

---

This essay served as the basis for a lecture Bush gave in February 1967 at the Pacific Service Center in Seattle, Washington. A somewhat altered version, under the same title, appeared in the volume *Science Is Not Enough* (New York: William Morrow & Co.,1967).

are being filled by those who have given the science, technique and practice of management formal study. But doctorates are still rare among its members, and likely to be accidental appendages rather than admission tickets among the elite.

The practice of management is taught, but not the art. Rarely it is, of course, when an inspiring teacher appears somewhere, but [the art of management] is not included in the curriculum. Yet success in management depends fully as much upon excellence in the art as it does upon the grasp of the mechanism of organizations, or the formal interrelations between men. It is this art that I wish to talk about, not to teach it, but to emphasize its central importance, if the profession is to merit recognition as a profession.

One can learn something of the art, not very much, by reading the story of success of great managers of the past. Not very much for several reasons. The scene changes, and what may have been artistic a decade or two ago might well be crude today. The writing of good biography is difficult; the writing of good autobiography is impossible. Well, not quite impossible; I have to make an exception for one such as [Harvard medical professor] Hans Zinsser [who was known for his literary writing and contributions to microbiology]; but he was not writing about management.

I suspect also that some of our finest managers do not quite know how they do it.

Before I proceed let me recite an incident, lest you misunderstand me. I would not have you think I am talking about some Machiavellian system of manipulating one's fellow men. Karl T. Compton was one of the most admirable, and most effective, leaders of men I ever knew. One day I went into his office and there on the table was a book, which was then receiving a great deal of ballyhoo, on the art of making friends and influencing people. I said, "Oh ho, are you reading that?" or something to that effect. This evidently jarred him; he did not reply, but [my question] stayed in his mind. Some minutes later, when there was a pause, he said, "The only true art of influencing people is complete absence of art." No truer word was ever spoken.

Now let us examine a case. The case system is now popular in teaching so let's follow it. Why was Horatio Nelson such a great leader? He understood the process of management certainly. His fleet was well provisioned, and his crews had been trained to shoot accurately. He knew how to delegate. Before Trafalgar he explained to all his admirals and captains his full plans. Collingwood was his second in command. He put him in charge of 12 ships, to cut off the enemy rear. Having done so he left him in complete control and did not interfere. But his success was not all due to these things.

Was it his slogans and signals, some of them seeming today a bit stilted? These are always emphasized in the histories. I do not think a tough group of seamen were much influenced by brave words.

A few days before the battle, the Admiralty in London ordered one of Nelson's admirals home for a court-martial. Note that the Admiralty once had an admiral shot on the deck of his own ship, for leaving the line of battle, even though he left it to attack the enemy. Nelson sent the admiral home all right, but he sent him in his own flagship, thus taking a ship out of the line just before meeting a nearly equal enemy. Why? We can say, of course, that if he lost the battle he was sunk anyway, and if he won the Admiralty would not dare to touch him. Quite so. But this was an act, not words, and the fleet understood: first that he was in command and would brook no interference by arm-chair admirals; second that he would use his full power to protect a subordinate against arbitrary injustice. No wonder he had a loyal fleet, and one that was bound to win. Nelson violated ancient, accepted rules for fleet combat, and won hands down.

The art of management has much in common with the art of painting. A painter can spend hours studying the mixing of color, the techniques of applying it to canvas, the suggestion rather than the depiction of form, and still never be an artist. Similarly a man can spend years studying organization charts, balance sheets and operating statements, even the use of computers and the queer ways of Madison Avenue, and never become a manager. Is the difference inborn, a matter of the sequence of nucleic acids on the chains of the chromosomes? I do not think so. Does the difference arise because one youngster is taught by his father that the world is his oyster and how to open it, while another is taught to love his neighbor? Well that can make a difference in adult character, no doubt, but I do not think it makes bad and good managers. I believe the difference, partly molded by these two things, depends rather upon the form that ambition takes, when young men come to the point where their logic, their observations and their will can build that ambition into the shape that will be theirs. And I believe it depends upon whether that ambition is to become a great leader of men, or just to make a success.

By this time you are probably thinking that I am drawing the contrast between hard and soft management, between tough and tender leadership. I am doing no such thing.

Certainly there are contrasting types. On the one hand is the arbitrary boss of whom all subordinates are scared stiff. At the other extreme is the manager who is so kind-hearted that he tolerates incompetence all about him. I have no use for either extreme. The dictator never has the benefit of skilled, honest counsel from his team; the easy-going manager never has a tight ship.

What I am saying, or trying to say, is that the technique of management may be learned from books, but the art of management must be learned from life. The art is far more important than the techniques. To learn it one must wish very much to do so. And one must be honest in the wish. . . .

Sometime after the war I went through a German mine, with the High Commissioner of Germany and the head of a mining company. When we passed a group of workmen they did not come to attention and click their heels, they kept on working. When we passed a workman in a narrow passage, he said, "Gruss auf" [greetings], and kept on going. No scout preceded us, yet I saw no man loafing. I concluded I was in a well-managed company. In many German plants I found the same general atmosphere. At the same time, in France I encountered surliness and soldiering on the job. I suspect this, rather than just economic factors, accounted for the fact that Germany recovered from the war fast.

These are just little details? So they are. The world is full of little details. The great decisions will never be made wisely unless the little details are in order. Of course I do not mean that the president of the company should prowl the plant looking for dirt or poor lighting. I do mean that, if he radiates a proper attitude, it will be transmitted downward throughout the organization, especially if he has a "sixth sense" which tells him the point at which the channel of transmission is being blocked.

I can tell you how to determine whether a really accomplished manager is in command of an organization. See what his subordinates really think. I do not mean whether he knows his costs, understands his market, keeps reasonable relations with his labor, has courage and judgement on new ventures [or] makes a profit. We assume all these criteria are met. If a manager who really excels is present a subordinate will know, when he is given a job, exactly what he is supposed to do, will know that he will be backed up and not interfered with. Moreover he will know that, if he makes a mistake, the boss will tell him so, but tell no one else, and will reason it out with him so that he will not make the same one again. He will know that if he has tough luck and gets into a jam, the boss will labor to get him out of it, and will tell no one that he is doing so, although it will soon be known throughout the shop. He will know that, if he makes too many mistakes, or if he is just not up to his job, he will be eased out, but that the boss will attempt to accomplish this with the minimum possible damage. Finally he will know that, if he puts forth his best efforts, sweats at the job, and it turns out badly, the boss will himself take responsibility for the failure, for that is the heart of sound delegation.

In such a shop you will witness a change of atmosphere when a bit of skullduggery appears. There will now be no temporizing, and there will be no doubt who is running the show. Everything can wait, but not wait long, until the unethical or illegal practice is ferreted out and eliminated. Nor will a foreman pick on the girls in his section if the boss finds it out. In some strange way he will know if evil is present, yet he will employ no spies, nor will he query an individual without the knowledge of his superior.

Will his people be afraid of him? Yes, in the best sense. In the same way that a boy should be afraid of his father, even if the father is fully kind and understanding. The boy once saw his father girded with righteous wrath when someone let the air out of his tires, and he has no wish to experiment along those lines. Similarly in a sound organization everyone knows the boss packs a punch, for they have seen him in action.

I could go on indefinitely and so could you. Let me make just one more point. An important part of the art of management is the art of listening. Some managers do not listen at all, even when they appear to do so. Some listen too much, but merely listen. No manager can be expert in all phases of modern management; he is fortunate if he can be really expert in one. He has to depend on his team in phases where he is not fully versed himself. One of the best managers I ever watched in action was a great listener. He paid attention. The proof that he was doing so came soon. He would ask a question showing he was actually three steps ahead of the advisor. Was this quickness of mind, an innate gift, not to be developed by attention and practice? Not at all; it merely showed that this particular man had acquired the art of listening. And he got good advice from alert advisors, straight from the shoulder.

Now just one more word before I close. Many of you are members of the profession of management. The hallmark of a true profession is ministry to the people, exercised with pride and dignity. It is best exemplified in the medical profession, where there is a sound relation between doctor and patient, and of course in an intense form in the clerical profession of religion. But it is always present, or there is no true profession. The profession of management is young, and those in it will determine its form as it matures. The days of the manager whose eye is solely on the profit statement, who cares nothing for the community in which he operates, or the welfare of those who labor, or the obligations to further the broad security and prosperity of his country, those days are numbered. There are examples still about, and there always will be, just as there will always be parasites who play games to extract dollars, and add none to the commonweal. But, as the profession matures, there will be, more and more, the creation of an accepted aristocracy of management, consisting of those who have earned and who receive the full respect of their fellows. There will be managers who can be tough on occasion, but who are always kind. They will have breadth, and recognition of their responsibilities as citizens and as fellow workers. Above all, they will be men who have learned the technique of management, but have proceeded beyond this and learned the art.

54

"ON THE DIFFICULTY IN VIETNAM" (1967)

Like many other leading figures of the greatest generation, who came to prominence during the American triumphs during the World War II, Bush found the Vietnam War confusing and troubling. Bush's criticism of the U.S. military's performance in an earlier Asian land war—on the Korean peninsula—might have prepared him for the "difficulty in Vietnam," but instead he seemed flummoxed. "I see no way out of the mess at the present time," he writes in a letter to a fellow traveler in previous campaigns to contain the global spread of Soviet influence. His words of frustration stand as a reminder of the limits of science-based innovations in deciding the outcome of military conflict.

"ON THE DIFFICULTY IN VIETNAM"

Dear Tracy:

The difficulty in Vietnam as I see it is, first, that we cannot win, and second, that we cannot get out. One of the usual ways of winning a war is to destroy the enemy's army in the field. This is usually done by breaking through his lines, cutting his communications and generally

Vannevar Bush to Tracy Vorhees, June 27, 1967 (Vannevar Bush Papers, Vietnam File, MIT Archives).

overwhelming him. We can hardly do this in Vietnam since the communications are not within our range, and there is no formal battle line through which to break. The second way of winning a war is to capture the enemy's center of operations, usually his capital city, and thus to paralyze his government and control it. We cannot do this in Vietnam for the center of operations lies in North Vietnam and is beyond our range. The third way of winning a war, and very unpleasant one, is to continue it until a revolt occurs among the enemy's civilian population, or mutiny occurs in the enemy ranks. That is to continue the war until the enemy folds internally. It can also occur under these conditions: by the enemy government becoming sufficiently discouraged so that it will cease to fight. It seems to me that, as the situation stands in Vietnam, that sort of war weariness is far more likely to occur with us than with the enemy.

I hence conclude that there is no way out of the present mess which lies in winning the war. I also think that it is very unlikely indeed that we can get out by negotiation. Both China and Russia are undoubtedly delighted to find us in the present mess, inclined to perpetuate it and certainly disinclined to help us out of our troubles. Loss of manpower due to casualties will have very little effect on the Oriental population. I also feel that there is no solution in escalating. Every time we send in more troops, North Vietnam can match it by sending in a few and the situation remains as before. Extending the war to include North Vietnam would not, in my opinion, lead to very little change in our situation. It would indeed render the war a far more serious one, more expensive in money and casualties. I do not think it would bring China into the war formally, but I do think we would then find ourselves in much the position we are now except that the base of supplies and control would lie across a border in China rather than across a border in North Vietnam. The situation that prevents us from terminating the war now would, as far as I can see, prevent our terminating it then. Some of our people have advocated that we just pull out as a way of terminating. To do so would mean all of our friends in South Vietnam would be promptly murdered. If we deserted our friends in this way, our influence throughout the world would rather largely disappear and deservedly so. It has also been suggested that we withdraw our troops into a perimeter around a selected part of the country in which our friends would be gathered. No war was ever won by remaining on the defensive. The line would be very long indeed and would take great bodies of troops to defend it. The enemy would have complete choice as to when and where to attack. I think any such withdrawal to a compound would probably leave us worse off than we are now.

In other words, I see no way out of the mess at the present time. But I do think that is what we ought to be thinking of, and I believe that the thinking people in this country, those who might find for us some mode of dignified escape, are largely thinking about it in the wrong terms. Or not thinking about it constructively at all. In fact, most of what we hear consists merely of statements that we must stop the war, without any suggestions whatever as to how to do it. I therefore hope that quite a few men of influence will, in any way they possibly can, state the present facts of the situation bluntly to the American public. And I would hope that out of this might come suggestions for a reasonable solution from some wise individual somewhere. . . .

Cordially yours,

V. Bush

55

"DO BIRDS SING FOR THE JOY OF SINGING?" (1970)

Bush enjoyed sailing off the coast of his beloved Cape Cod, a geographic cape extending into the Atlantic Ocean from the southeastern corner of mainland Massachusetts. The grandson of two sea captains, Bush expressed pride in the sailing activities of his ancestors and often invoked the activity of navigating ocean waters as a metaphor for how he faced difficult decisions in his own life with mental toughness and resilience. Along with sailing, Bush took casual pleasure in the appreciation of nature, especially bird life. In this short reflection, Bush demonstrates that while his world consisted largely of building and managing complex technological systems, he appreciated the majesty and mystery of the natural world. Here he asks, "Do birds sing for the joy of singing," and answers in the affirmative because, he reasons, why else would "the complexity of their songs" be "far greater than is needed for recognition or for marking of reserved areas." Jerome Wiesner, the science adviser to President Kennedy, was so taken by Bush's words on birds that he quoted them in his memorial essay for the National Academy of Science following Bush's death in 1974.

"DO BIRDS SING FOR THE JOY OF SINGING?"

For the title of this book, I have drawn on the wealth of the vocabulary of the youth of our times. Theirs is a pungent stock of words, and action marks most of them. In my time, it has been my good fortune to have a piece of the action here and there in varied circumstances. It has been a pleasant experience for me to review some of the more rugged of these, and some of the more serene.

Do birds sing for the joy of singing? I believe they do. The complexity of their songs is far greater than is needed for recognition or for marking of reserved areas. I have become acquainted with a catbird who obviously derives pleasure as he tries out little phrases on his own. Moreover, I believe that evolution produced birdsongs, and the joy that goes with them, because of the survival value they bestow.

He who struggles with joy in his heart struggles the more keenly because of that joy. Gloom dulls, and blunts the attack. We are not the first to face problems, and as we face them, we can hold our heads high. In such spirit was this book written.

Appeared as the foreword to Bush's late-life memoir, *Pieces of the Action*, dated January 31, 1970; reprinted in Jerome B. Wiesner's essay on Bush, published in the National Academy of Sciences, *Biographical Memoirs: Volume 50* (Washington, DC: National Academies Press, 1979).

56

THE REVOLUTION IN MACHINES TO REDUCE MENTAL DRUDGERY (1970)

Five years before the birth of the first personal computers in 1975, and four years before his death in 1974, Bush published his final reflections on how human-built devices, placed on the top of a desk, can enhance an individual's cognition, or purposeful thought. This essay, drawn from Pieces of the Action, *provides a fitting closing selection to this volume of Bush's essential writings because, as he admits here, "the idea" of mind-expansion and cognitive enhancement, brought about by computers, was "almost constantly" on his mind in the final years of his life. In anticipating the personal computer and the spread of electronic information networks, Bush identifies the tension between augmenting, or enhancing, human thought through computation and the process of replacing, or substituting, human thought with "artificial intelligence." Here Bush substantially extends his forecasts on the "whole revolution . . . of machines to do man's mental drudgery." Importantly for historians of computing and contemporary designers, Bush provides valuable insights into his memex concept on the expansion of human memory through electronic means, and his prototype for a "rapid selector," which he imagined as "not much bigger than a typewriter." Both the memex and the rapid selector illustrate how Bush anticipated broad outlines of the effects of computers on human consciousness and culture. Bush was also among the first Americans to write in clear, concise, and sometimes elegant language about the revolution in information engendered by digital devices and networks. Significantly, his insights widely influenced such computer pioneers as Douglas Engelbart, inventor of the computer pointing device, or mouse, Ted Nelson, creator of hyperlinks, and Google founders Sergey Brin and Larry Page.*

THE REVOLUTION IN MACHINES TO REDUCE MENTAL DRUDGERY

There are several somewhat distinct aspects of the whole revolution being brought about by the employment of machines to do man's mental drudgery for him, a development as significant for his future welfare as was the introduction of machines to perform his physical labor.

One such aspect is the handling of data, and this breaks into two parts. First, there is the storing and retrieval of figures and coded instructions. The modern computer has extensive memories for this purpose, and consults them very rapidly indeed. Second is the storing and finding of letters, sheets of figures, diagrams, all the complex records on which the conduct of business and libraries depends. This second phase of data handling has moved forward relatively slowly but is now speeding up.

Just before World War II, I got into this affair and, with the late John Howard, at M.I.T. built what we called a rapid selector. It had a somewhat strange history. I include it here because it illustrates well the type of invention which is almost inevitable once the technical elements it combines have advanced to the point where they are adequate for the purpose.

The way it worked was this. A reel of movie film had photographed on it a mass of data, perhaps 200,000 frames, each of some sort of document. The edge of this record film had a set of transparent dots on a black background which coded the adjacent frame. One set up the code of an item to be searched for by depressing a number of keys. As the roll of record film progressed through the machine, every time the set code coincided with the code of a frame, a group of photocells triggered a flashing lamp and that item was photographed onto a new strip of film—the reproduction film. Thus one could run through the reel and receive promptly a reproduction of every item in the collection called for

Excerpted from Bush's *Pieces of the Action*, pages 187–192.

by the set code. To accomplish this several things were necessary. First, the record film had to move rapidly as it left one reel and was wound up on another. This was easy. In a movie camera the film moves at only 24 frames per second. But it has to stop at each frame for an exposure and then start again. With continuous motion—no stops—the film could readily travel 40 times as fast.

Second, when a desired frame was located, a photograph of it had to be made, without stopping the film, and without blur. Harold Edgerton [an electrical engineer and M.I.T. professor who invented the stroboscope to make stop-action and high-speed, multiple-action possible] had solved this problem. His flash lamp gave intense light pulses of very brief duration. Exposures could be made in micro-seconds instead of centiseconds as in an ordinary camera. Thus no blur occurred in a picture of a fast-moving film.

Finally, the coding had to be worked out so that only desired items would be selected and photographed. For this good sensitive photocells were then available. Several methods of coding were developed. The one easiest to describe, though not finally used, was as follows: Opposite each frame in the long film were the transparent dots, say a hundred of them, arranged in groups of 10 each. At a keyboard one punched out the code of the item desired, producing a small card with 10 holes punched in it, arranged in a pattern according to the keys that had been pressed. The card was inserted in the selector so that the fast film ran close under it and was strongly illuminated. As the record film moved, light passed through the card and the film and impinged on a photocell whenever a hole in the card and a dot in the film registered in position. If nine or fewer such coincidences occurred, the photocells remained inactive, paid no attention. But if there were 10 such coincidences, indicating that the frame then in position corresponded exactly to the impressed code, the photocell triggered the flash lamp to take a picture. Since, at the exact instant this occurred, the chosen frame was not in a position to be easily photographed, a delay circuit was introduced and the flash lamp fired when the fast-moving film had advanced two frames, to bring the chosen frame in front of the camera lens. The camera was an ordinary movie camera, with its shutter always wide open. Every time the flash lamp operated, it advanced the sensitive reproduction film one frame. There was some tricky gadgetry, which need not be described, to catch chosen frames that happened to be close together on the record film, and there was a rig so designed that, after a run, the short piece of exposed reproduction film, containing only a few frames, could readily be cut off and developed.

Suppose one had a roll of film containing a million pictures of checks, all duly coded, and [one] wanted to find a check made out for $1,036.48. One would punch this number on the keyboard, adding zeroes to fill out the ten places. One would then run the film through, taking about 16 minutes (or less if one stopped as soon as the camera clicked), would have a picture of the sought check. One could get several pictures if there were coincidences in amounts.

Of course the coding would usually be more complicated than this. For a library, for example, the individual frames of the record film might be summaries of technical articles, coded in accordance with the subject matter, author, date, and so on.

The whole device was not much bigger than a typewriter. Of course it was crude compared to the equipment that can be built today. It was an early step along a path that has now become elaborate indeed. That step had no more than been taken when the war came along, and several things happened. First, the patent application on the device, with its assignment to M.I.T., somehow got lost or not followed up. Second, the machine itself was taken over by the group working on enemy codes and that was the last I saw of it. One does not worry much about long-range planning problems during a war.

After the war the Department of Commerce wanted such machines to handle some of its data. No company could be found to introduce the device into commercial use, for there was no patent to protect them. Commerce finally had one or two made and put to work. But an art like this moves rapidly, and there were soon better machines available. Today there are a wide variety of ways of going about this job of finding a needle in a literary haystack. With magnetic tapes, transistors, printed circuits, there is a whole family of machines for handling data. The Eastman Kodak Company has made an especially interesting one called Minicard. It combines the sorting features of the old punched-card machines with the searching ability of the rapid selector. It is not very fast as yet, but no doubt soon will be.

All this is a long way from a girl digging through a file cabinet. Someday libraries will be fully mechanized. Then without leaving one's office, it will be possible to pick up the phone, dial in a code and have the actual paper one is looking for almost instantly at hand. Something of the sort has got to happen or our libraries will become buried in the mass of books and articles now being printed, and searching in the old way will become hopeless.

Something more than 30 years ago, pondering these latter problems in the light of the work in the selector as well as the analytical machines, I conceived the idea of a machine that should be an extension of the personal memory and body of knowledge belonging to an individual, and should work in a fashion analogous to the working of a human brain—by association rather than by categorical classification. I called the device a memex, and published a discussion of it in the *Atlantic Monthly* in July 1945. Essentially, a memex is a filing system, a repository of information and a scheme of searching and speedily finding a desired piece of information. It utilizes miniaturization, high-speed photography, memory cores such as computers embody, and provisions for the coding of items for recall, the linking of code to code to form trails, and then the refinement or abandonment of trails by the machine as it learns about them. It is an extended, physical supplement for man's mind, and seeks to emulate his mind in its associative linking of items of information and their retrieval as a result.

No memex could have been built when that article appeared. In the quarter-century since then, the idea has been with me almost constantly, and I have watched new developments in electronics, physics, chemistry and logic to see how they might help to bring it to reality. That day is not yet here but has come far closer as I set out in an essay, "Memex Revisited," published in 1967 [collected in his book of essays entitled, *Science is Not Enough*].

The heart of the idea is that of associative indexing whereby a particular item is caused to select another at once and automatically. The user of the machine, as he feeds items into it, ties them together by coding to form trails. For the usual method of retrieving an item from storage we use a process of proceeding from sub-class to sub-class. Thus in consulting a dictionary or an index, we follow the first letter, then the second and so on. The rapid selector worked this way. Practically all data retrieval in the great computers follow this method.

The brain and the memex operate on an entirely different basis. With an item in consciousness, or before one, another allied item is suggested and the brain or the memex almost instantly jumps to the second item, which suggests a third and so on. Thus there are built up trails of association in the memory, of brain or machine. These trails bifurcate, cross other trails, become very complex. If not used they fade out; if much used they become emphasized. Thus a desired item may be found far more rapidly than by use of a clumsy index.

Millions of items are stored in man's brain, memories, sheets of data. Suppose I wish to recall what Aunt Susie, whom I haven't seen for 20 years, looked like. I don't start by turning to all of the pictures in my mind or storage where the name begins with S, and so on. Not at all. My brain runs rapidly—so rapidly I do not fully recognize that the process is going on in some cases—over when I saw her, what was the occasion, what were her mannerisms and suddenly her picture is before my mind's eye. The goal of a memex is comparable: the use of the associative trail. Although we cannot hope to equal the speed and flexibility with which the mind follows an associative trail, it should be possible to beat the mind decisively in the permanence and clarity of items resurrected from storage.

Here is where the ability of the digital computer to learn from its own experience . . . comes into play. Suppose that a memex has as its master a mechanical engineer and that he has a trail which he uses very frequently on the whole subject of heat transfer. The memex "notices" (we have to use such terms; there are no others) that nearly every time he pursues the trail there are a series of items on which he hardly pauses. It takes them out of the main trail and appends them as a side trail. It also notices that when he comes to a certain item he usually goes off on a side trail, so it proceeds to incorporate this in the main trail.

[The memex] can do more than this; it can build trails for its master. Say he suddenly becomes interested in the diffusion of hydrogen through steel at high temperatures, and he has no trail on it. Memex can work when he is not there. So he gives it instructions to search, furnishing the trail codes likely to have pertinent material. All night memex plods on, at ten or more pages a second. Whenever it finds the words "hydrogen" and "diffusion" in the same item, it links that item into a new trail. In the morning its master reviews the new trail, discarding most of the items, and joining the new trail to a pertinent position.

ACKNOWLEDGMENTS

I have spent the past thirty years studying the life, mind, and times of Vannevar Bush, and many people have helped me on my journey. For the opportunity to assemble this collection of Bush's writings for new generations of students, scholars, citizens, and policymakers, my greatest thanks, and largest debt of gratitude, goes to Dr. Richard D. Bush, Van's oldest son. Dr. Bush, a medical doctor who lived in Belmont, Massachusetts, agreed with me that I should undertake the process of selecting and editing a volume of his father's writings.

Following the publication of the biography I wrote, *Endless Frontier: Vannevar Bush, Engineer of the American Century*, Dr. Bush encouraged me, in a letter, to "bring back into print some" of his father's published and unpublished work in order to promote a better understanding of the thinking of Vannevar Bush on a range of topics of historical, cultural, literary, technological, and scientific interest. Specifically, Dr. Bush, as his father's literary executor, asked me, as his father's sole biographer, "to promote public awareness" of his father's legacy and to highlight "the role of science policy in American life."

I'm profoundly grateful for Richard's support and I am sorry he is not with us today to appreciate the results of a task that took far longer to complete than I anticipated.

I owe an equal, if different, debt to Columbia University Press. The press supported my project despite epic delays in delivering a manuscript.

Life got in the way (and not Bush's life, but my own). At Columbia, my editor, Stephen Wesley, helped me through the editorial process with grace, style, and sound judgements. Frustratingly, I delivered a manuscript in the midst of a pandemic that upended Columbia's normal operations, yet despite these disruptions to normal routine, Stephen deftly guided my manuscript through the process of review and refinement. He is the reason the book exists, and I salute him and his Columbia colleagues.

Many scholars in history, science policy, and social studies of technology nourished and sustained me over many years, and I can only mention here a few people who helped on this edited volume. Louis Galambos, one of my personal heroes in the field of history and for decades the editor of Dwight Eisenhower's papers, advised me on the value of including my own commentary on Bush's writings. Martin Sherwin, the outstanding historian of the nuclear age, made suggestions on relevant selections and commentary. Paul Ceruzzi, a titan among historians of computing, reinforced my belief that Van Bush's thinking on the future of computing deserved equal place with his political and managerial writings on innovation, international security, and the relationship between science, engineering, and American life.

At Arizona State University, where I served as a professor of practice for ten years, from 2010 to 2020, Dr. Michael Crow, the university's president, provided an invitation, inspiration, stability, guidance, and support. Dr. Sethuraman Panchanathan, for many years the chief of ASU's research enterprise, shared with me his own passion for Vannevar Bush and his enthusiasm for particular selections. "Panch," who became the director of the National Science Foundation (NSF) in 2020, reinforced my belief in the timeliness and utility of this book for leaders in research funding and administration. Jason Lloyd, an editor in ASU's Washington, DC office, helped me make critical decisions at the earliest stage of this project. I'm deeply grateful for Jay's assistance. Among ASU's outstanding faculty, I especially benefited from conversations on science policy with Dr. Robert Cook-Deegan (who also generously shared documents with me), Dan Sarewitz, and Mahmud Farooque. Other ASU faculty who assisted me: the biologist James Collins; the historian of fire, Stephen Pyne; Clark Miller and Jameson Whetmore, of the School for the Future of Innovation; the intellectual historian Hava Samuelson; the historian of Livermore Labs, Sybil Francis, with whom I taught a history of nuclear weapons class for

ASU undergraduates; and Benjamin Hurlbut and Gaymon Bennett, intellectual polymaths who nourished my curiosity and nudged me in the direction of greater clarity and depth.

Van Bush's ties (historic and symbolic) to those who direct our nation's science policies and research administration inevitably brought me into their orbit, however tangentially. Bush remains the touchstone for every contemporary effort to ground research spending and goals in the public interest, and as both a figure in history and a symbol of the possibilities of publicly funded science, many contemporary actors seek to imagine what Bush might think or do in our current situations. I'm drawn to these inheritors of Van Bush's world, because they carry on in his grand tradition of public service. I wish to especially thank the outgoing director of the NSF, France Cordova, an outstanding physicist and research administrator. Dr. Cordova graciously invited me to deliver the keynote address about Vannevar Bush at an NSF gathering in February 2020, in order to mark the seventy-fifth anniversary of Bush's call for the creation of a national research foundation, which led to the creation of the NSF. At the event, I saw Neal Lane for the first time in many years. As both a former presidential science adviser and an NSF director during the Clinton, Dr. Lane possesses a resume that comes close to matching that of Van Bush in terms of service to the American nation in the years after the breakup of the Soviet Union and the end of the Cold War. Dr. Lane kindly contributed a foreword to the book and brought a piece of Bush's writings to my attention.

Many archivists have helped me generously since I began studying Bush's papers more than thirty years ago. For this volume, specifically, I received significant assistance from Chamisa Redmond, a senior information specialist at the Library of Congress in Washington, DC.

Finally, I wish to thank John Markoff, author and journalist for the *New York Times*, whose writings on the technological present and past set the highest standard for insight and accessibility. Throughout the project, John shared with me his vast knowledge of technological history, and his calm, cool demeanor. This volume is dedicated to him. My wife, Constance Okon, deserves my deepest appreciation for her sympathy, understanding, assistance, and sense of perspective. My debt to her cannot be tallied.

INDEX

342 INDEX

Printed in the USA
CPSIA information can be obtained
at www.ICGtesting.com
JSHW021435221024
72172JS00002B/7